AF235203

WHY DON'T RABBITS RULE THE WORLD?

First published in Great Britain in 2026 by Cassell, an imprint of Octopus Publishing Group Ltd
Carmelite House
50 Victoria Embankment
London EC4Y 0DZ
www.octopusbooks.co.uk

An Hachette UK Company
www.hachette.co.uk

The authorized representative in the EEA is Hachette Ireland, 8 Castlecourt Centre, Dublin 15, D15 XTP3, Ireland (email: info@hbgi.ie)

Text copyright © Ben Sparks and Dr Katie Steckles 2026
Design and layout copyright © Octopus Publishing Group Ltd 2026
Pop art copyright © Joe McDermott 2026 (pages 3, 4, 7, 14, 33, 38, 44, 45, 46, 50, 74, 78, 91, 94, 111, 114, 118, 127 (top), 138, 145 and 165)

Distributed in the US by Hachette Book Group
1290 Avenue of the Americas, 4th and 5th Floors
New York, NY 10104

Distributed in Canada by Canadian Manda Group
664 Annette St., Toronto, Ontario, Canada M6S 2C8

All rights reserved. No part of this work may be reproduced or utilized in any form or by any means, electronic or mechanical, including photocopying, recording or by any information storage and retrieval system, without the prior written permission of the publisher.

Ben Sparks and Dr Katie Steckles assert the moral right to be identified as the authors of this work.

ISBN: 978-1-78840-599-7
eISBN: 978-1-78840-620-8

A CIP catalogue record for this book is available from the British Library.

Printed and bound in China.

10 9 8 7 6 5 4 3 2 1

Commissioned by: Stephanie Selcuk-Frank
Editor: Scarlet Furness
Copy Editor: Sam Hartburn
Art Director: Yasia Williams
Designer: Nikki Ellis
Illustrations (excluding pop art): Nikki Ellis and Sarah Fisher
Pop Art Illustrator: Joe McDermott
Assistant Production Manager: Lisa Pinnell

MIX
Paper | Supporting responsible forestry
FSC
www.fsc.org
FSC® C008047

WHY DON'T RABBITS RULE THE WORLD?

DR KATIE STECKLES & BEN SPARKS

MATHEMATICAL ANSWERS TO LIFE'S KEY QUESTIONS

CONTENTS

INTRODUCTION

In the modern world, it's increasingly apparent how much every aspect of our world uses, and is described by, mathematics. Mathematics has always been there, and its beauty and utility have been appreciated by many for thousands of years.

The way we use mathematics in science, technology and communications gives huge power and insight to those who understand, even at a basic level, how our mathematical world functions.

Mathematics has always been about seeking to understand the world, as well as exploiting its patterns and insights. This book is a chance to explore some obvious, and some not-so-obvious, mathematical connections.

In Section 1 we'll explore mathematics which applies to social and romantic situations, while in Section 2 we examine the mathematics involved in cheating, deception and truth – always important in a world worried about misinformation.

Section 3 tackles the way your money works (or doesn't), and includes some important reminders on investment and financial risk-taking.

In Section 4 our mathematical lens focuses on the way animals, plants, and the natural and physical world can be modelled mathematically to give surprising and useful insights – while Section 5 comes back to our very human inclination to play and compete against each other, and examines the mathematical principles behind sports and games.

Dotted throughout the book are also some puzzles and paradoxes, which we hope are interesting enough to take time to ponder before reading on to the conclusions.

Enjoy the journey!
Ben and Katie

LOVE AND RELATION- SHIPS

1

WHY DO YOUR FRIENDS HAVE MORE FRIENDS THAN YOU?

Picture this: you're at an event where you don't know anyone. You check your phone, and see a feed full of posts about parties, social gatherings and holidays that you weren't invited to. You can't help but wonder why everyone seems to have more friends than you do. This odd and slightly disheartening observation isn't all in your head – it's a well-documented phenomenon called the **friendship paradox**.

First described by sociologist Scott L. Feld in 1991, the friendship paradox is the counterintuitive observation that, on average, **your friends have more friends than you do**. In 2012, researchers in America studying social networks found that the average person on Facebook had 245 friends, but the average number of friends each of those friends had was 359.

This doesn't mean that every single one of your friends is more popular than you – but rather, if you lined up five of your friends, and took an average of how many friends they had, it would likely be greater than the number you have. And while you might feel like this is somehow your fault, it isn't: it's just a mathematical fact, which has some wide-reaching implications – from transport networks to the COVID-19 pandemic.

WHAT IS THE FRIENDSHIP PARADOX?

We can think of a social network (either online, or the real networks people form) as a collection of points (or vertices), joined by lines (or edges). This is a mathematical structure known as a **graph** – and graphs allow us to analyse many different types of structures and relationships.

For example, we could define a graph, denoted G, as:

- a set of vertices (also called **nodes**), V, each corresponding to a person in the social network, and
- a set of edges, E, each one representing a friendship – if two people are friends, their vertices are joined by an edge.

THE WHOLE GRAPH: $\quad G = (V, E)$

THE SET OF VERTICES: $\quad V = \{\text{Alice, Bob, Carol, David}\}$

THE SET OF EDGES: $\quad E = \{(\text{Alice, Bob}), (\text{Bob, Carol}),$
$\quad\quad\quad\quad\quad\quad\quad\quad (\text{Bob, David}), (\text{Carol, David})\}$

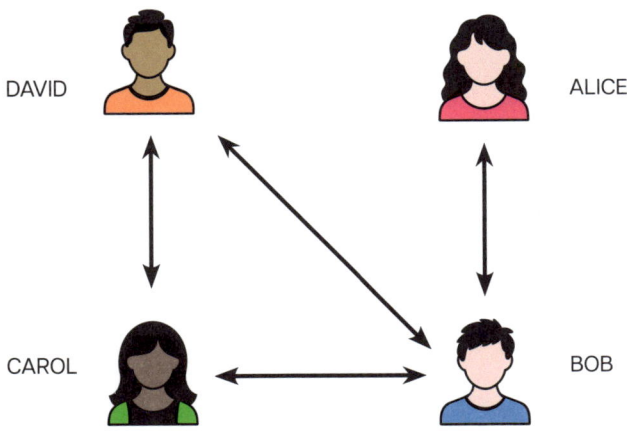

The graph shows us that Alice has only 1 friend, Bob. Bob is the popular one, with 3 friends. David and Carol have 2 friends each. In this network, the average person has 2 friends – so, nothing strange yet.

PERSON	FRIENDS
ALICE	1
BOB	3
CAROL	2
DAVID	2
AVERAGE	2

To observe the friendship paradox, we need to consider how many friends a person's friends have. For Alice, Bob is her only friend – so the average number of friends her friends have is 3. Out of Bob's 3 friends, who have 1, 2, and 2 friends each, the average number of friends is 1.67. For Carol and David, who each have a friend with 2 friends and another with 3, this is 2.5.

PERSON	FRIENDS	AVG FRIENDS OF FRIENDS
ALICE	1	3
BOB	3	1.67
CAROL	2	2.5
DAVID	2	2.5

So for three out of the four people here, the average number of friends that their friends have is higher than their own number of friends. In this case, Bob is the only person this is not true of, since he has a lot of friends – but in general, it's the case for the majority of people in a given network.

This is how networks behave in general – your individual experience is distorted by the effect of being part of a larger connected group. The paradox doesn't always work – it won't happen in certain extreme situations, such as when everyone has the same number of friends – but these kinds of situations are unlikely to occur in real life.

In real situations, there's also an element of bias – people will seek out more popular people to be friends with, since they are likely to be a more attractive friendship prospect (because they're more generous, or rich, or kind) – so the effect is amplified by popular people getting more popular. The result is that everyone else is left with this paradoxical feeling of unpopularity. But it will happen to almost all of us!

APPLICATIONS AND IMPLICATIONS

We can apply the mathematical ideas behind graph theory to many other situations involving connected networks. In particular, determining which vertices in the graph have the largest number of edges connected to them allows us to tackle real-world problems effectively.

- **PUBLIC HEALTH:** During an outbreak, individuals with many social connections are more likely to contract and spread infectious diseases. Public health strategies can leverage this by targeting vaccinations or interventions at these well-connected individuals, effectively slowing disease transmission.

- **MARKETING:** Focusing on influencers and individuals with extensive networks helps to promote products and ideas. By tapping into the networks of these key individuals, marketers can amplify their reach and increase the likelihood of a campaign going viral.

- **TRANSPORT:** Enhancing facilities, improving connectivity, and increasing train frequency on routes containing more well-connected train stations can enhance the overall travel experience for many passengers. (And the friendship paradox isn't specific to human relationships – within the Eurail train and ferry network, each city is connected to an average of 2.7 cities, but their neighbours are connected to an average of 3.8 cities!)

SO...

...why do your friends have more friends than you do? In this case, it's not you, it's maths.

WHEN SHOULD YOU STOP LOOKING FOR 'THE ONE'?

If you've ever spent any time dating, you'll know it's not exactly a formulaic process. It's messy, uncertain and full of second-guessing. You meet someone and think, 'They're pretty great, but...what if there's someone better?' The classic dilemma: stay and risk wondering 'What if?', or move on and risk regretting it.

At the heart of this is a familiar question with surprising depth: when is the optimal moment to stop searching and commit? Too soon, and you might miss out. Too late, and all the good ones may be gone.

OPTIMAL STOPPING

This can be modelled mathematically as an **optimal stopping problem**. This is where we're faced with a series of options to choose between, presented one at a time. As we consider each option we have the choice to say 'yes' or 'no'. But once we say 'no', that option is gone forever: we can't go back.

Originally introduced in the context of hiring a member of staff (and sometimes called the 'secretary problem'), this model applies to many scenarios: choosing a parking space, buying a house or – in our case – looking for a partner.

It's worth noting that, in real-world hiring situations, employers often interview everyone first and then contact their chosen applicant afterwards. In contrast, the model assumes we make an immediate 'yes or no' decision after each option, which makes it more appropriate for situations like dating.

To make this work, we need to assume three things:

1 Some options will be better than others: Whether we're hiring or dating, quality varies. For our model, we need to define a way to compare two candidates, and determine which one is better.

2 Options arrive in a random order: If we knew they were arriving in decreasing order of quality, we could just pick the first one, so we're assuming the ordering is random.

3 We know how many options we have to choose between: Imagine we have a fixed amount of time to find the perfect partner, and we date a new person every week, so we know what the total number of prospects will be.

Step 1: Sample without choosing

The first step to this method is to observe the first few options to 'get a feel' for what's out there. People often do this intuitively when they view potential houses to buy, or try different restaurants: the early ones are for comparison, not commitment.

The initial phase helps you calibrate. You build a sense of what a good candidate looks like, which helps you spot someone exceptional later.

Step 2: Choose the first one who's better

Once that sampling phase is over, the next step is simple: pick the first person who's better than all the ones you've seen so far.

For example, imagine you start by reviewing (dating) five candidates. This is probably enough to establish a baseline – now all you need to do is look for the next candidate which beats all of those. Of the first five you saw, you should have been able to pick a favourite by comparing them. For each new candidate, you can compare them to your current favourite – as soon as you get someone better, you go with that.

This method has been shown to be the most likely way to find the best possible candidate from your range of options. The only question then is, when should you shift from sampling to actually choosing?

THE 37% RULE

Optimal stopping theory tells us that the way to make sure the one you end up with is as good as possible is to try out around 37% of the possible candidates – a little over a third – then pick the next one after that which beats all the options so far.

This value of 37% comes from a mathematical constant called e. This number is approximately equal to 2.71828, and it's often used to define an **exponential function**: $y = e^x$. Here, x is the input value to the function, and we find the output y by raising the number e to the power of x. We could use any value in place of e here, but choosing this particular value gives a function – shown in the graph – which has a slope of exactly 1 when $x = 0$: the line is pointing diagonally upwards at 45° when it crosses the y-axis.

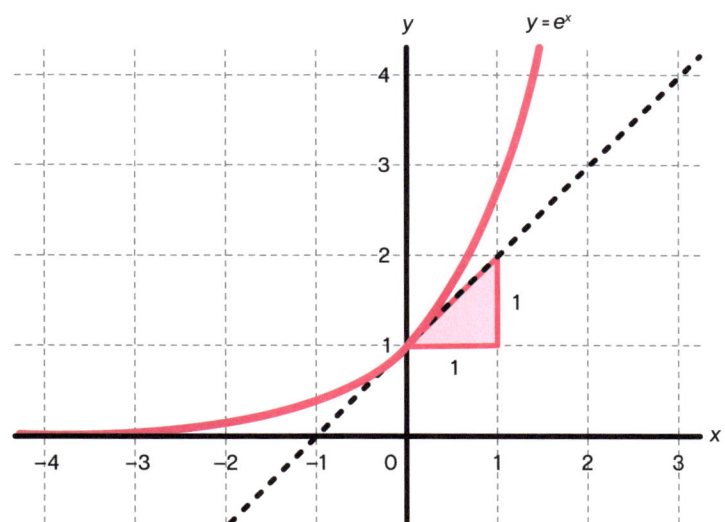

This function is important in studying exponential growth – when the rate at which something is growing gets bigger the more of it there is. This kind of growth often occurs in situations like bacterial replication, or the spread of fires. The **exponential constant** e is an important value in mathematics, because the function e^x describes this type of growth. (For more on exponential growth, see page 86.)

And e is useful elsewhere too – our 37% comes from dividing the number of total candidates by e – and since $1/e \sim 0.37$, this gives us about 37%. The method is likely to give you a reasonable result overall, but it will find you the overall best candidate, from the selection given, around 37% of the time.

In the example given earlier, we might know there are 14 potential dates lined up. Then finding 37% of 14 tells us we'd want to review 5 before we start considering options for 'the one'.

WILL THIS HELP YOU FIND LOVE?

Of course, this doesn't guarantee you'll get the best option: if the absolute best candidate is among the first few you see, you'll miss out, since they'll be within your calibration set. But assuming things are randomly distributed, this is most likely to help you find 'the one' (that's the best of the current available dating pool). Then you just need to hope your second date goes well!

PLAN YOUR PARTY WITH PERMUTATIONS!

Hooray – you're organizing a party! But who do you invite? Do your guests know each other? Will you need to make introductions? Will everyone get along? And most importantly...will there be pizza?

Whether it's counting handshakes or sending a team to collect food, and whether you know it or not, you're deep in the realm of **combinatorics** – the branch of maths concerned with counting, choosing and arranging things.

HOW MANY HANDSHAKES?

Suppose everyone at your party shakes hands with everyone else. How many handshakes occur?

This obviously depends on how many people are present at the party.

If there are only two people, you have one handshake (and a decidedly anticlimactic, if intimate, party).

3 PEOPLE: 3 HANDSHAKES (1–2, 1–3, 2–3)
4 PEOPLE: 6 HANDSHAKES (1–2, 1–3, 1–4, 2–3, 2–4, 3–4)
5 PEOPLE: 10 HANDSHAKES (1–2, 1–3, 1–4, 1–5, 2–3, 2–4, 2–5, 3–4, 3–5, 4–5)

What's going on here? This is one of the most everyday examples you might encounter of a **non-linear sequence** – the number of handshakes is not increasing by the same amount each time. If you drew a graph of this sequence it would not be a straight line – hence the term non-linear.

With this kind of sequence, it can be hard to spot a pattern. You might notice that the amount by which the sequence is increasing is itself increasing by the same amount each time (in this case, by 1). This is a characteristic of a **second-order relationship**, often referred to as a **quadratic relationship**.

Trying to draw a diagram often helps with understanding. In this case we can draw a graph (see page 11) – and the question becomes the same as asking 'How many edges do you need to join every vertex to every other vertex?'

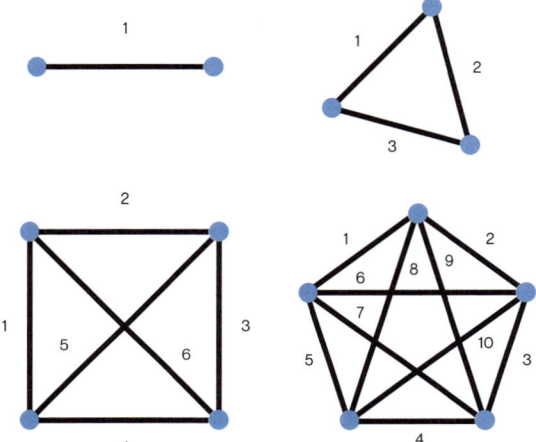

The edges needed to join 2, 3, 4 and 5 vertices

This will get harder to draw as the numbers get bigger, but at least we have some visual confirmation for the numbers so far.

Alternatively, we can break down our counting. Assume that one person starts off the handshaking by greeting everyone else. If there are 6 people present, the first person will engage in 5 handshakes (one for every guest other than themselves).

The next person has already greeted the first guest, so they have 4 left to greet. It might become clear that the third handshaker has only 3 to do, the next 2, and finally one handshake between the fifth guest and the remaining ungreeted guest.

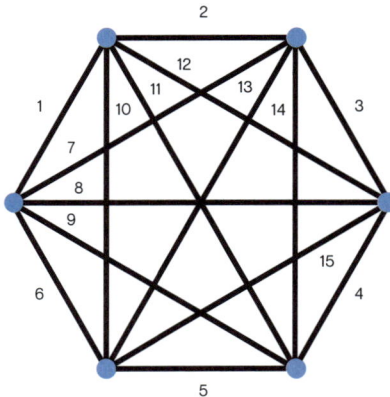

The edges needed to join 6 vertices

$5+4+3+2+1=15$. The diagram backs this up – although it is now getting harder to count the lines.

This sequence of numbers is called the **triangular numbers**, because for each number of handshakes, we can form a triangle made from that many objects.

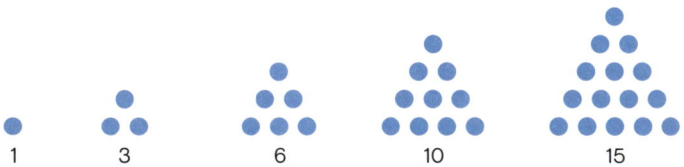

1   3   6   10   15

The first five triangular numbers

This reflects the way we counted handshakes, by adding together smaller and smaller numbers – e.g. 15 is 5 (the bottom row) plus 4 (the next row up) plus 3, plus 2, plus 1.

In general, if you have n people, there will be $\frac{n(n-1)}{2}$ handshakes. This checks out for 6 people: let $n = 6$, and our formula becomes $\frac{6 \times 5}{2}$ or 15, as expected. In fact, this is half the area of an $n \times (n-1)$ rectangle of dots, and half the rectangle is a triangle.

A mathematician might refer to this outcome as 'n choose 2' (sometimes written as $^nC_2$), because it counts the ways of choosing **two** things from a list of n things.

WHEN ORDER MATTERS (AND WHEN IT DOESN'T)

If Alice shakes hands with Bob, that's the same event as Bob shaking hands with Alice. The order doesn't matter. There's only one handshake between them, not two. But in some situations, order does matter. Imagine instead that Alice punches Bob (hopefully not at your party). That's a very different scenario from Bob punching Alice. In this case. 'A punches B' is not the same as 'B punches A'. So now we do care about who does what.

This difference is the starting point for what mathematicians refer to as **combinations** and **permutations**, and associated ways of counting them.

Choosing a pizza team: when order doesn't matter

Imagine you had a sudden need to send out 3 people from the party of 30 people, perhaps to pick up a pizza order. Choosing Fred, Wilma and Barney is the same outcome as choosing Barney, Wilma and Fred. You don't mind who goes first, second or third.

A selection of members where the order doesn't matter is sometimes called a **combination**.

Listing all the possible teams there could be is frustratingly long-winded, but it can be stated mathematically as '**n choose 3**' (or $^nC_3$) – as in, how many combinations of 3 things can be chosen from n things (where the order doesn't matter).

In this case, where $n = 30$, the answer is 4,060 combinations.

Ranking party guests: when order does matter

Now, suppose you're picking the top three most interesting people at the party (awkward, but let's go with it). This time, the order matters. Fred first, Wilma second and Barney third is different from Barney first, Wilma second and Fred third.

This is what we call a **permutation**: counting how many different ordered arrangements of 3 people there are from a group of 30.

Mathematically, this is described as **'n permute 3'** (or $^nP_3$), or 'how many ways are there of picking **3** things from n things (where the order I pick them matters)'.

In this situation, where $n = 30$, the answer is 24,360.

That's 6 times the number of combinations – because there are 6 different ways to rearrange any group of three people. We'll give some more explicit ways to calculate these numbers on the next page.

Counting those orders!

We could also write the number of ways to order three people as 3! – here, the exclamation mark means 'factorial', meaning to multiply this number by all the positive whole numbers below it.

Counting the number of ways of ordering a list is an important part of these discussions, and that's exactly what this 'factorial' calculation does.

For example, to count the orderings of 3 people, there are 3 choices for the first place, then only 2 choices for second place, and whoever is left is in third place:

$3 \times 2 \times 1 = 6$

To order 7 people, it's the same idea, There are 7 choices for first place, 6 choices left for second place, 5 left for third place and so on:

$7 \times 6 \times 5 \times 4 \times 3 \times 2 \times 1 = 5,040 = 7!$

SUMMARY OF FORMULAE

For each of these different ideas, we have different formulae describing how to calculate the total number of possibilities.

Type	Question	Formula	Example
Ordering	How many different ways are there of ordering a list of size n?	$n!$	How many ways of putting 6 people in a line? $6 \times 5 \times 4 \times 3 \times 2 = 6! = 720$
Combinations	How many ways are there to choose r things from a list of size n, where order doesn't matter?	'n choose r' or $\binom{n}{r}$ or ^nC_r $$\frac{n!}{r!\,(n-r)!}$$	How many different 5-card poker hands are available from a standard pack of 52 cards? '52 choose 5' or 2,598,960 possible 5-card hands.
Permutations	How many ways are there to choose r things from a list of size n, where order does matter?	'n permute r' or ^nP_r or $$\frac{n!}{(n-r)!}$$	How many different playlists of 40 ordered songs could you choose from a library of 300 songs? (The order matters, as every DJ knows!) '300 permute 40' = approx. 8 with 97 zeros after it (more than the number of particles in the observable universe – these lists get big!)

You may notice that the permutations formula is very similar to the combinations formula, but has just been multiplied by $r!$ – this is because it scales up by precisely how many ways there are of reordering the r things.

In summary, any question you ask about counting things, choosing things, or putting things in order is likely to involve these three big ideas: combinations, permutations and ordering. These come under the general name of **combinatorics**. So next time you're planning a party, make sure you correctly combine your cocktails and perfectly permute your pizzas!

THE SECRETS BEHIND ONLINE DATING

Ever wondered how a dating app claims to 'find your match'?

Different apps have different philosophies. Some match people by mutual interests, or by personality traits – and others calculate 'compatibility scores'. While we can't tell you about any secret proprietary algorithms, there are definitely some ways to establish mathematically how closely matched you might be to someone else.

GETTING THE DATING DATA

Imagine a simple dating questionnaire:

	Strongly disagree	Disagree	Neither agree nor disagree	Agree	Strongly agree
Score	1	2	3	4	5
A I love being outdoors	☐	☐	☐	☐	☐
B Being kind is more important than being right	☐	☐	☐	☐	☐
C I enjoy in-depth discussions of technical detail	☐	☐	☐	☐	☐
D I love football	☐	☐	☐	☐	☐
E I like to have a lazy day with nothing planned	☐	☐	☐	☐	☐
F I love cars and engines	☐	☐	☐	☐	☐
G The way something looks matters more than the way something works	☐	☐	☐	☐	☐
H I like cooking	☐	☐	☐	☐	☐
I I am always ready to take a risk	☐	☐	☐	☐	☐
J I enjoy reading books	☐	☐	☐	☐	☐
Total score:					

If two people answer this questionnaire, how do we measure how similar their responses are?

To answer this, we need to work out how **far away** one set of answers is from another set of answers.

INTRODUCING DISTANCE METRICS

The word 'metric' comes from the same origin as the word 'measure' – just like the unit of distance: the metre. But a **distance metric** refers to a more general idea of 'closeness' between mathematical objects. In this case the objects are lists of 10 numerical answers to the questionnaire. A list of numbers like this is often referred to as a **vector**.

Let's start small. Here are the sample responses from four people, for question A. Using the language of vectors, we have 4 vectors, and each one has length 1.

	A
AMOS	4
BRYONY	3
CANDACE	4
DESMOND	1

Here, Amos and Candace answered the same, so we could say that they're similar or 'close'. But what if we add question B? Now each response is a vector with length 2, or a two-dimensional (2D) vector.

	A	B
AMOS	4	3
BRYONY	3	4
CANDACE	4	2
DESMOND	1	3

Now it's less obvious which pairs of respondees are 'close' in their answers – some match on one question but differ on the other. There are several options for how we could measure this. Here's one option:

- Calculate the differences in corresponding question answers; lower values mean they're closer. Add these differences together to get an overall 'closeness'

Comparing three of the possible pairings, by this metric, Amos and Candace are closer than Amos and Bryony, who in turn are closer than Candace and Desmond.

	DIFFERENCE IN QU. A	DIFFERENCE IN QU. B	TOTAL 'DISTANCE'
AMOS & BRYONY	1	1	2
AMOS & CANDACE	0	1	1
CANDACE & DESMOND	3	1	4

Here's another, more visual option for a metric: put the results on a coordinate grid, treating answers to question A as an x-coordinate and answers to B as a y-coordinate.

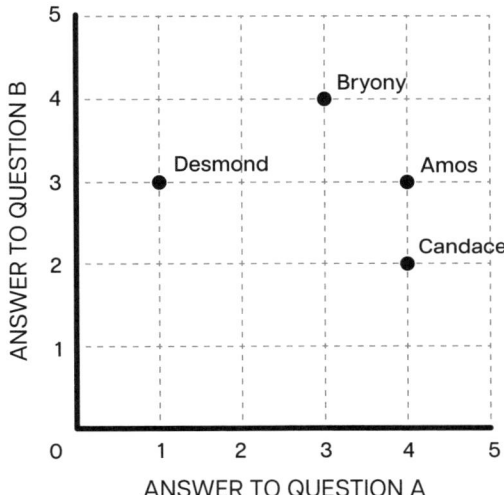

The graph gives a more intuitive reaction to the 'how close' question. Amos and Candace are a distance of 1 apart, the distance between Bryony and Amos is between 1 and 2, and the distance between Bryony and Desmond is just over 2. This time it actually *is* a distance between two points in space. To work out how far apart two points are, we need the famous-from-school-but-when-will-I-use-it-fact known as **Pythagoras' theorem**.

For example, we can calculate the distance between Bryony at point $(3, 4)$ and Desmond at point $(1, 3)$.

This involves finding the difference in the scores (which are the perpendicular sides of the triangle joining the points), squaring them, adding them, and then square rooting to find the hypotenuse – which in this case is the distance between the points. This is called the 'Euclidean metric' – although some call it a 'Pythagorean metric' because it's using Pythagoras' famous theorem.

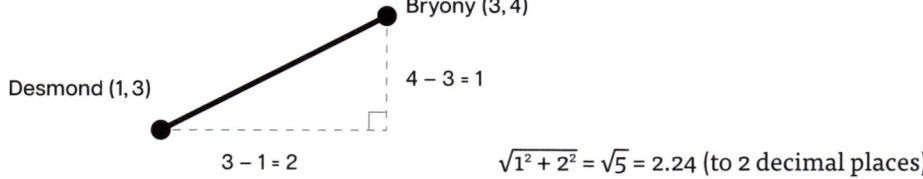

Desmond (1, 3)

Bryony (3, 4)

$4 - 3 = 1$

$3 - 1 = 2$

$\sqrt{1^2 + 2^2} = \sqrt{5} = 2.24$ (to 2 decimal places)

You can also use this diagram to visualize the first distance metric we discussed, using just the difference in the score. It's the length of the route between two points if you only travel horizontally or vertically.

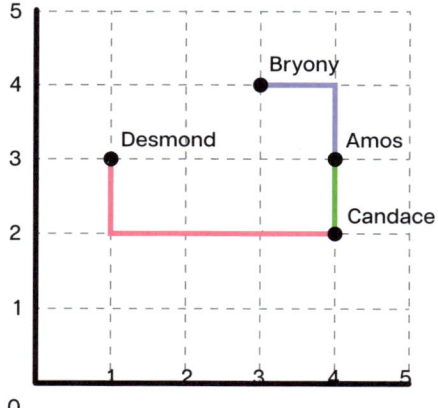

Distance from Bryony to Amos = 2
Distance from Amos to Candace = 1
Distance from Desmond to Candace = 4

This metric is sometimes called the **taxicab metric**, as if you were in a taxi in a city built in a grid system – imagine a yellow taxicab in New York, restricted to driving along the roads.

These examples show us that there's more than one answer to the question 'how close are these two things' – and we've only used the first two questions from our quiz so far. For now, we'll continue with the Euclidean metric (which in 2D is the Pythagorean straight-line distance between points).

	A	B	C
AMOS	4	3	3
BRYONY	3	4	5
CANDACE	4	2	5
DESMOND	1	3	1

If we use the answers to three questions, we can do exactly the same visualization, except this time we need three coordinates, and a 3D plot to represent our vectors, which are now three-dimensional.

It's harder to see the three-dimensionality on a printed page, but you can imagine each person being assigned a point inside a cube, with their coordinate triple being their three answers to the three questions.

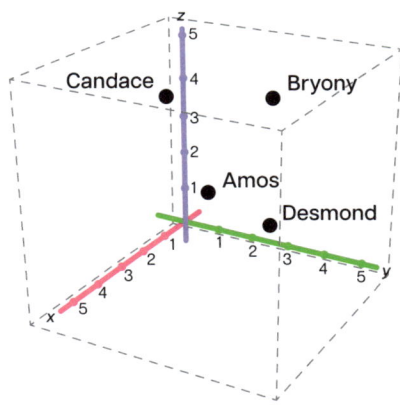

Responses for questions A, B and C plotted in three dimensions – here Desmond = (1, 3, 1) and Bryony = (3, 4, 5)

The answer to 'how close are these two points?' can be answered in exactly the same way as before – Pythagoras' theorem works for 3D diagonals too.

By visualizing Desmond (1, 3, 1) and Bryony (3, 4, 5) at the corners of a cuboid, Pythagoras theorem can also be used in three dimensions.

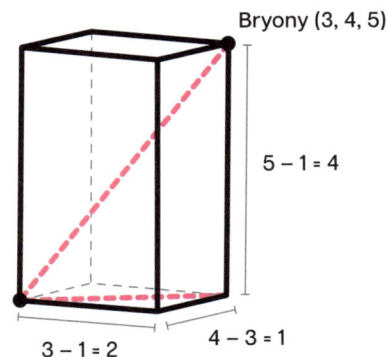

Distance $= \sqrt{2^2 + 1^2 + 4^2} = \sqrt{21} = 4.58$ (to 2 decimal places)

Based on these answers, the closest matches are Candace and Bryony, and Candace and Amos (they're the same distance apart).

So far so good: but no self-respecting dating app is going to use only three questions, and it's not obvious how to visualize all this with data from four questions or more. Visualizing a four- or five- or six-dimensional plot is not something that we find intuitive – since we live in a three-dimensional universe, it's hard to imagine going above that.

Here's where mathematics really shows its power – we can just carry on anyway. Even though it's hard to picture, the idea of Euclidean distance works just as well in any higher number of dimensions. So we can apply the same rules to our 10-question data and get some 'match measures' which would be very hard to see at a glance.

	A	B	C	D	E	F	G	H	I	J
AMOS	4	3	3	2	4	2	3	2	1	4
BRYONY	3	4	5	3	3	3	2	4	2	5
CANDACE	4	2	5	1	4	3	4	5	4	3
DESMOND	1	3	1	5	4	3	3	5	4	1

For example, the Euclidean distance from Bryony to Desmond now is:

$$\sqrt{(3-1)^2+(4-3)^2+(5-1)^2+(3-5)^2+(3-4)^2+(3-3)^2+(2-3)^2+(4-5)^2+(2-4)^2+(5-1)^2}$$

$= \sqrt{48}$

$= 6.93$ (to 2 decimal places)

(And, incidentally, this is not the 'furthest apart' pairing in the mix.)

No matter how many questions you want to incorporate, we just need to calculate the sum of the squares of all the differences between the two people's answers.

Taking the square root of the answer gives the Euclidean distance between any two people's answer sets, and this gives one option for a 'compatibility score'. The lower the score, the better the match. In this (completely fictional) group, based on these (completely fictional) answers, an 'Amos–Bryony' pairing is more likely to succeed than any other, even though they didn't answer exactly the same on any question.

At this stage you have the power to tweak the details as you wish. For example, if someone's answer to question B is more important to you than any of the others, you could give it more importance in the calculation by multiplying it by a 'weighting' factor.

It's probably best not to rely on this particular questionnaire to find your own dating matches, but this principle is almost certainly used behind the scenes in some form on any matching app you might use.

SO...

...how closely matched is your date? One way to measure it is to calculate the n-dimensional Euclidean distance between your answer vector and theirs, using Pythagoras' theorem.

If telling your date this doesn't impress them, please don't blame us!

ARE YOU MATHEMATICALLY UGLY?

Take a look at yourself in a mirror. Would you say you're conventionally attractive? They say beauty is in the eye of the beholder. But can we use mathematics to create a more objective and logical rating system for attractiveness? Some people have claimed to be able to...

MIRROR, MIRROR

Humans are part of the natural world, and subject to all the same processes of evolution and development as any living thing. We have a fundamental code that describes how we'll grow: DNA. The patterns encoded in our genes give instructions to our cells about how to grow and divide, which proteins to produce and roughly what we'll look like. But an individual's journey from a single cell to a full organism involves plenty of complex interacting influences – including genetic mutations – which can affect how we turn out.

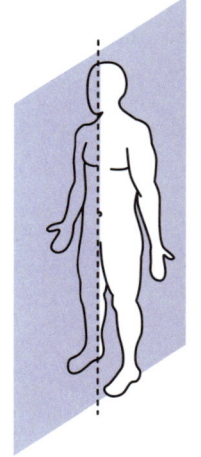

The theoretical blueprint for a human – as for many types of living creature – has a form of mathematical perfection to it: **bilateral symmetry**. This means it should be possible to draw a line straight down the middle of a person, from the top of their head to the bottom of their feet, and have everything be perfectly mirrored either side of the line. It's actually not a line, but a plane – called the **sagittal plane**, it's like a flat sheet running down through the body and from front to back, with everything reflected in it like a mirror.

This works for plenty of other creatures too. With few exceptions, mammals, birds, fish and even insects exhibit this type of symmetry – along with many of other kinds of natural structures, like leaves, and even certain flowers, such as orchids.

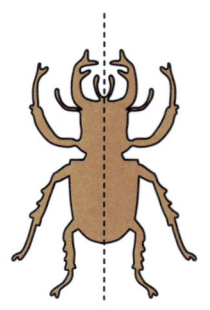

Symmetry has several advantages biologically: it reduces the complexity of the organism, since you have to store only half as much information, and can repeat the same arrangement on both sides. It also provides advantages in movement: being able to walk and run in a coordinated way is made easier if your limbs are symmetrical.

Roughly 99% of organisms display this kind of symmetry, although there is the odd exception. A small number of birds have beaks that curve to one side, and certain types of fish have both eyes on one side, since they spend most of their time lying on the seafloor. Some animals have non-symmetrical shells, like snails whose shells are spiral-shaped on one side, and a narwhal's tusk is helical. There are also some species of owl with ears that are differently positioned on each side, which lets them pinpoint the location of prey more accurately.

And not every part of a human body is symmetrical: the internal organs have an asymmetrical arrangement, with the heart on the left side of the body and the liver on the right. But, externally, the human body shape is (theoretically) perfectly symmetrical.

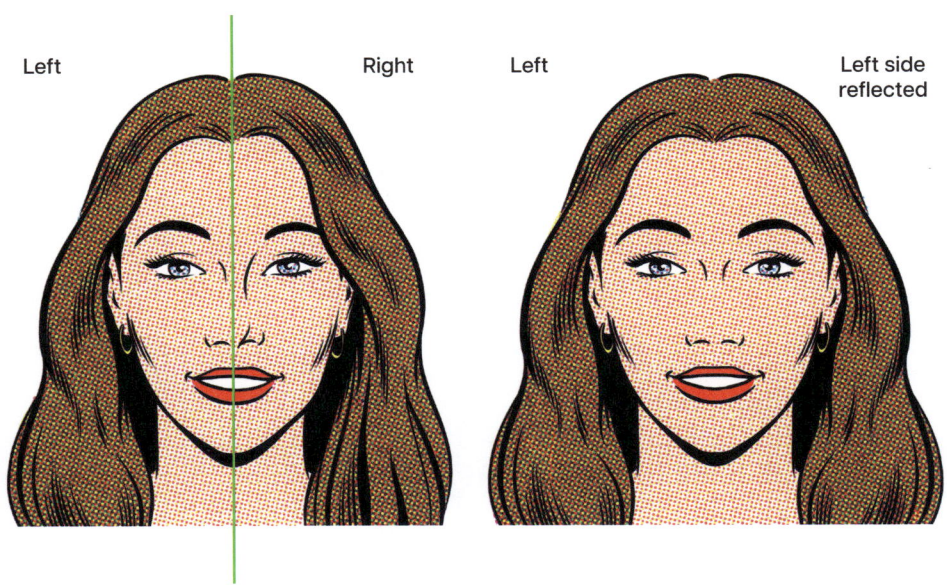

Left   Right   Left   Left side reflected

Comparing a face (left) to an image made from two copies of the same side (right), reflected, we can see how close to symmetrical their face is

In practice, development of things like facial features is subject to minor mutations and environmental factors, meaning that human faces are rarely perfectly symmetrical. But if you examine the faces of people considered to be 'beautiful', they are often faces exhibiting more perfect symmetry.

And it makes sense from an evolutionary perspective that this is a trait we'd find attractive: the more factors you've been subjected to that might modify your development, the less high-quality you could be considered as a mate, since you're more likely to have been exposed to damaging environments or adverse evolutionary pressures.

So when you're staring at your own face in the mirror, compare the two halves – do they match up? Don't worry if you're not perfectly symmetrical: this is completely normal, and if anything it means you've had a more interesting life!

THE MOST BEAUTIFUL NUMBER

There's another famous mathematical concept commonly associated with beauty: the **golden ratio**. To define this, we need a little mathematics.

The **Fibonacci numbers** are a sequence of numbers that starts with 1, then another 1, and from there it follows a simple rule: each number in the sequence is the sum of the two before it. This is why we need to start by defining two terms: we can then add them to get the next term, which is 2, then $1+2=3$, and $2+3=5$ and so on.

The golden ratio can be found by looking at the ratios between these numbers: for example, the ratio between the second and third terms of the sequence is $2/1 = 2$. Then $3/2 = 1.5$, and $5/3 = 1.666...$ The ratio is different for each pair of numbers, but as we continue along the sequence, the ratios start to get closer and closer to a particular number. Mathematicians would say the ratios **converge** (for more on convergence, see page 148).

The place they're heading is a particular number that can't be written as a simple ratio of two whole numbers.

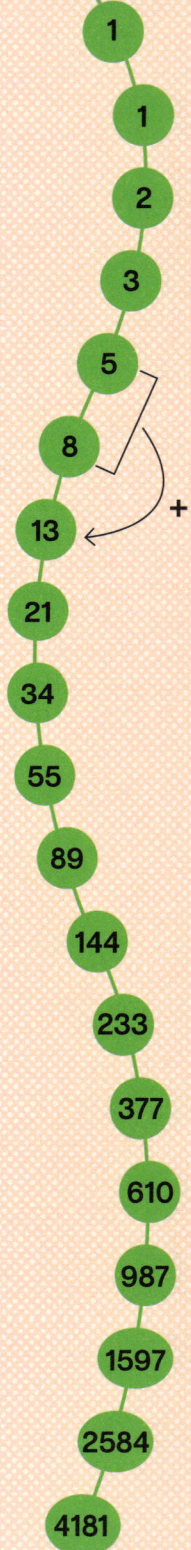

1	2	3	5	8	13	21	34	55	89
1/1	2/1	3/2	5/3	8/5	13/8	21/13	34/21	55/34	89/55
=1.0000	=2.0000	=1.5000	=1.6667	=1.6000	=1.6250	=1.6154...	=1.6190...	=1.6177...	=1.6182...

It's approximately 1.618..., and can be precisely written as $\frac{\sqrt{5}+1}{2}$. Since this expression contains a square root (of a non-square number), the number on the top isn't a whole number, and the decimal expansion of the ratio goes on forever (just like the decimal expansion of π, and any other irrational number). This means it isn't the actual ratio of any pair of consecutive Fibonacci numbers, no matter how far along the sequence you go: it's the value the ratios are getting closer to, but will never reach.

φ = 1.6180339887498948482045868343656381177203091798057628621354486227052604628189024497072072041893911374847 5...

Sometimes denoted using the Greek letter phi (φ), the golden ratio is this imaginary limit: the ratio at the 'end' of an infinite list of ratios (which, being infinite, doesn't have an end). And as a number, it has many interesting and beautiful properties.

For example, calculating $1/\varphi$ gives the same answer as subtracting 1 from φ: the answer starts with 0.618... (with the same numbers after the decimal point as the original). This is the only number that this works for (technically, it also works for one other number, but it's derived from the same ratio: $-1/\varphi$). You can also square φ, and get $\varphi^2 = \varphi + 1$; again, it's the only number that does this.

$$\frac{1}{\varphi} = \varphi - 1$$

$$1 = \varphi^2 + \varphi$$

$$\varphi^2 - \varphi - 1 = 0$$

$$\varphi = \frac{\sqrt{5}+1}{2}$$

This 'golden ratio' also crops up in the ratios between certain lengths on geometric shapes – since it contains a $\sqrt{5}$, it unsurprisingly has connections to pentagons and five-pointed stars, and appears in certain fractals.

Since the golden ratio has beautiful mathematical properties, it has been suggested by some that it has connections to other forms of beauty: the 19th-century German psychologist Adolf Zeising wrote a whole book explaining why this beautiful ratio was deeply connected to external physical human beauty.

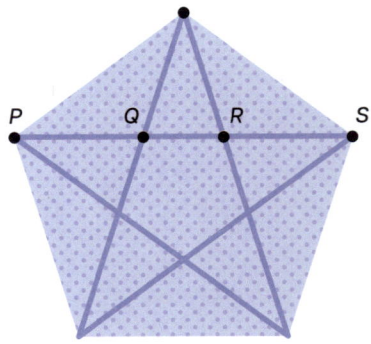

$PS/PR = PR/PQ = PQ/QR = 1.618...$

He claimed that the ratio between a person's height and the height of their navel (belly button) should be the golden ratio; he also found many pairs of lengths on a person's face which he claimed would be in the golden ratio if the face was truly beautiful.

There have also been many claims about people using this ratio to produce beautiful art and elegant engineering, including the paintings of Leonardo da Vinci, and pieces of important historical architecture like the Parthenon in Athens or the pyramids of Giza – it's claimed that if you measure certain lengths on these and compare their ratios, you'll find the golden ratio there too.

Unfortunately, a lot of these claims are likely wishful thinking: there's no evidence that the ancient Egyptians even knew about the golden ratio, and the measurements that are claimed to match this ratio so perfectly often involve drawing some suspiciously thick lines to allow for a bit of fudging.

It's also not clear that the golden ratio has anything to do with human beauty: the scientists claiming this seemed to base most of their work on specific types of (usually European) faces, ignoring the fact that a lot of people, who are definitely considered to be beautiful by many, have very different facial proportions.

And if you take a shape as complex as the human body and start comparing measurements, you're bound to find some

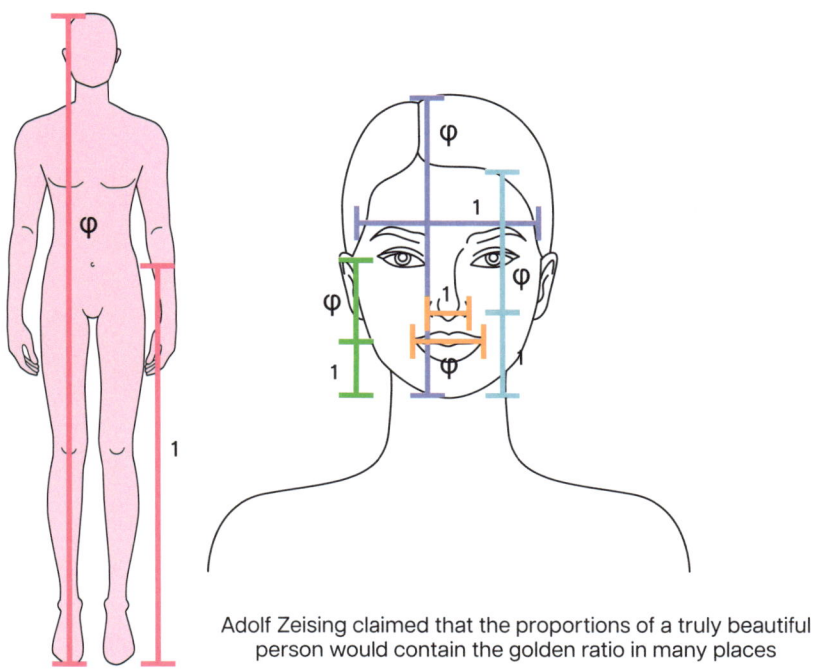

Adolf Zeising claimed that the proportions of a truly beautiful person would contain the golden ratio in many places

things that are about 1.6 times as big as each other – but this varies hugely between people, due to genetic variation, and nobody's been able to back up any of these claims with actual evidence that they correlate with attractiveness.

Luckily, though, the golden ratio retains its mathematical beauty – and there are situations where its properties do mean it crops up in the natural world. For more about the golden ratio and how it can make plants and flowers both beautiful and efficient, turn to page 137.

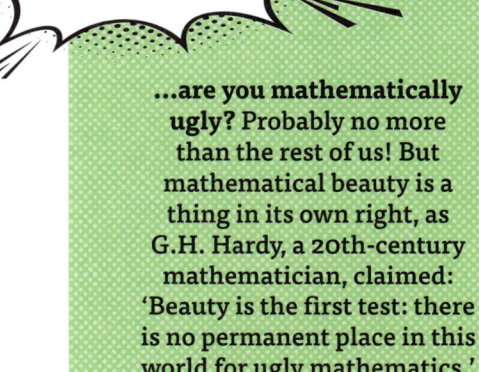

SO...

...are you mathematically ugly? Probably no more than the rest of us! But mathematical beauty is a thing in its own right, as G.H. Hardy, a 20th-century mathematician, claimed: 'Beauty is the first test: there is no permanent place in this world for ugly mathematics.'

WHEN SHOULD YOU PLAN A JOINT BIRTHDAY PARTY?

Have you ever met someone with the same birthday as you? We're just talking about the day and month here; don't worry about the year. It might have felt like a cool coincidence – maybe even cause for a joint party.

But how rare is that really? Well, if (like one of the authors of this book) you're an identical twin, it's a lot more likely than average.

But for most people, we can think of this question as a probability puzzle:

How many people would you need to gather together to be confident that there would be a shared birthday between at least one pair at that gathering?

Like any question of this sort, the detail is important. Here the word 'confident' is not very well defined, so let's assume we mean: 'expect it to happen more often than not', which you may recognize as a shorthand for 'greater than 50% probability', or 'better-than-even odds'.

This has become a famous question in the study of mathematical probability, because the answer, even with totally reasonable assumptions, is wildly different to what most people's intuition tells them.

Jot down your own guess before reading on.

THE BIRTHDAY PROBLEM

Let's start with some easily established extreme cases.

- With 1 person, there is a 0% chance of a shared birthday, as there's no-one for them to share it with.

- With 367 people it's 100% certain that a pair of shared birthdays exists, because there are only 366 possible options (including 29 February). Even if you manage to gather 366 people, all with different birthdays, the next arrival would be guaranteed to share a birthday with someone already in the room.

THE PIGEONHOLE PRINCIPLE

Mathematicians call this form of reasoning 'the pigeonhole principle', after the idea of filling up possible options like postboxes or pigeonholes. (See page 110 for more about the pigeonhole principle.)

Many people, when pushed, will get to this stage and estimate the answer for a 50% chance to be in the region of 183 people. You might have already guessed that this is wrong – but just how wrong it is can be cause for surprise.

You need only 23 people to have a better-than-even chance of there being at least one shared birthday pair.

THE CALCULATIONS

The probabilities for different numbers of people can be calculated step-by-step. We'll show the calculations in a spreadsheet alongside the explanations.

With probability questions it's often easier to think about the opposite situation: **what are the chances that no-one present shares a birthday?**

We can work our way up from the first case: 1 person. As we've already noted, there's a 100% chance they won't share a birthday with anyone else, since there's no-one there to share it with.

	A	B	C
1	No. of people	Chance of no shared birthday	Chance of at least one shared birthday
2	1	100.00%	
3	2	=364/365	
4	3		
5	4		

For two people, we need the second person to 'dodge' the first person's birthday. If we assume all birthdays are equally likely (and ignore leap years for now) this would happen about $\frac{364}{365}$ (or 99.7%) of the time.

	A	B	C
1	No. of people	Chance of no shared birthday	Chance of at least one shared birthday
2	1	100.00%	
3	2	99.73%	
4	3	=B3* 363/365	
5	4		

For three people not to share, we need the first two people in the room not to share (which we just calculated the chance of), and then the third to dodge the two birthdays already claimed. There's a $\frac{363}{365}$ (99.4%) chance of this. We're assuming the events are independent, so we can now multiply these probabilities together: $\frac{364}{365} \times \frac{363}{365} = 99.2\%$.

	A	B	C
1	No. of people	Chance of no shared birthday	Chance of at least one shared birthday
2	1	100.00%	
3	2	99.73%	
4	3	99.18%	
5	4	=B4*(365 – A4)/365	

We can continue in this way, multiplying each previous result by the chance of the next arrival dodging all birthdays already claimed – for n people that would be $\frac{365(n-1)}{365}$. This is the sort of calculation spreadsheets are excellent at.

	A	B	C
1	No. of people	Chance of no shared birthday	Chance of at least one shared birthday
2	1	100.00%	=100% – B2
3	2	99.73%	
4	3	99.18%	
5	4	98.36%	

All that remains is to recognize that we are calculating the opposite of what we're after (the chance of no shared birthday), so the actual percentages we want are 100% minus these numbers.

	A	B	C
1	No. of people	Chance of no shared birthday	Chance of at least one shared birthday
2	19	62.09%	37.91%
3	20	58.86%	41.14%
4	21	55.63%	44.37%
5	22	52.43%	47.57%
6	23	49.27%	50.73%
7	24	46.17%	53.83%

Generating the lists of these calculations in a spreadsheet reveals that the probability we're after (the chance of at least one shared birthday) increases quite quickly.

50% is reached on the 23rd row

What's even more surprising is how quickly the chances become close to 100%. By the time you've got 61 people in a room, there's a 99.5% chance of winning a bet that there'll be at least one pair of shared birthdays.

The assumptions we've made (all birth dates are equally likely, and independent of each other) probably make this an underestimate of the true probability. Birthdays are not equally distributed across the year, and this makes the chance of a match in general even higher.

WHY WE'RE BAD AT GUESSING THIS

One of the authors has frequently put this to the test, betting their own hard-earned money that there'll be a birthday match in samples (of around 60 people) from live show audiences. Offering stakes of £50 against a sceptical audience member's £1 has earned them a satisfying income over the years (even though small and of questionable ethicality). The £50 offer is tempting enough to lure a player in, it seems – but if they knew the true odds of winning, the audience member should've been holding out for a potential prize of around £200, compared to their own £1 stake.

So why does this feel unintuitive to so many people?

One possible reason is that we tend to frame the question from only our own perspective: 'what's the chance someone shares *my* birthday?'

That's a completely different question. For each person in a room with n people, there's a $\frac{364}{365}$ chance you don't share their birthday, so the probability that one of them *does* share your birthday is $1 - \left(\frac{364}{365}\right)^{n-1}$.

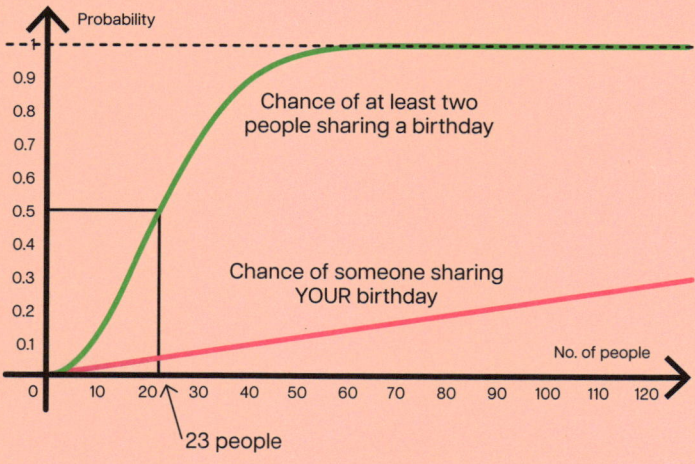

This gives a much smaller probability than the original birthday problem – which is about *any* pair sharing a birthday.

The birthday problem may seem surprising because we underestimate how fast the number of possible pairs in a gathering grow. This is the same calculation seen in the 'how many handshakes' question (see page 19). For a gathering of n people, the number of different pairs is 'n choose 2' or $^nC_2$, which grows much quicker than n, meaning that as the value of n increases, the chance of a shared birthday increases dramatically too.

n	$^nC_2$
10	45
23	253
30	435
60	1770

IS THIS ACTUALLY IMPORTANT?

There are deeper things to learn from this. While most commonly framed around parties, the 'birthday problem' can be applied to more important issues like computer security.

Suppose you're generating random passwords. If the total number of possible passwords is large, you might think it's unlikely that two people will get the same one. But once enough people sign up, the chance of a collision (two people being given the same password) rises surprisingly fast – just like our birthday example. In the high-stakes world of computer security this sort of thing moves quickly from the territory of 'curious fact' to 'multi-million-dollar-security-risk'.

This is just a glimpse into how non-linearity can lead our intuition astray (see the car crash problem on page 186 for another non-linear counter intuitive situation). Feel free to test this problem out at the next large gathering you're at – but don't blame us if you lose money on ill-advised social bets.

THE POISONED PIZZA PUZZLE

Learning to think strategically, out-play your opponents and come out on top are important skills in many different situations. Whether it's a fun game or a difficult business negotiation, logical decision-making is crucial – and mathematical thinking can give you an advantage.

One slice of the pizza is poisoned; can you avoid taking that slice?

Here's a game that will test your strategic-thinking skills.

A pizza is cut into thirteen equal slices. Two people plan to share the pizza, but they both know that one of the slices contains a lethal quantity of poison (and they know the one with the olive on it, it's ugh). They've agreed to take it in turns picking up slices of pizza – each person can choose to pick up one, two or three slices on their turn, to put on their own plate.

If you had to play this deadly but delicious game, what would be your strategy? Would you want to pick first or second, and how many slices would you take?

Try the game now – find a friend and draw thirteen dots on a page, then take it in turns to cross out one, two or three dots and see if you can force your friend to take the last one.

This game is a variation on a historical game called **Nim**, which is thought to be the oldest game in the world. While it has many different variations, they all have some common features: there are two players picking from a pile, and on each turn they can choose from a limited range of options for how many to take.

Some variations involve the winner being the player who takes the last item, and others – like our poisoned pizza – are set up so the winner is the player who can force their opponent to take the last one.

FINDING A WINNING STRATEGY

In the case where a turn involves taking one, two or three items, we can find a strategy that guarantees we win. To understand the strategy it helps to work backwards from the end of the game, as follows.

- If you leave your opponent with one slice left they have to take it. (Opponent has 1 slice = YOU WIN)

- To get to a situation where there's only one slice left, you need to take all but one of the slices.

- You can do this if there are either two, three or four slices left (since you can take one, two or three).

- So if your opponent had five pieces left to choose from, they would be forced to put you in one of those three situations, since they'd take either one, two or three. (Opponent has 5 slices = YOU WIN)

If there are five slices left, your opponent must leave you with either two, three or four slices, so you can force them to take the last slice

Now the goal is to leave your opponent with five slices. Reducing five slices to one slice is possible at any stage of the game because, whatever number your opponent picks, you can choose a number to make it up to a total of four slices removed. So, if you left nine slices, you could let your opponent choose whichever number they like, and then get the total to five. (Opponent has 9 slices = YOU WIN.) And similarly, from thirteen slices you can force the total to be nine. (Opponent has 13 slices = YOU WIN.)

In this game starting with thirteen slices, all you need to do is allow your opponent to choose first, and then whatever they take, you can get the number back to one more than a multiple of four, continuing until you leave the one slice at the end.

For different variations on the game, this strategy can be tweaked. If you wanted to be the player who took the last piece you would need there to be exactly a multiple of four left – in that case, you could create and maintain that by going first and taking one slice; then you'd be on a multiple of four (twelve pieces), and could force the total to go to eight, then four, then zero.

Once you understand the rules of a game like this, you can use similar strategies to give yourself an advantage. Other variations of Nim involve taking things from multiple piles, different rules about the number of objects you can take and different numbers of players – but they can all be beaten using a clever strategy.

COMBINATORIAL GAMES

Nim is an example of a particular class of games called combinatorial games: these usually involve two players, who take turns to make a choice. They also require everyone to have access to all the relevant information at all times – so it can't involve players having a hidden hand of cards, or anything where one player knows something the other doesn't.

The category includes games like Chess, and the ancient Chinese board game Go, which involves taking it in turns to place black and white stones on the squares of a grid and trying to 'capture' areas of the board. It also includes simple games like Noughts and Crosses – also known as Tic-Tac-Toe – which are small enough that they can be analysed by writing out every possible move at each stage.

Mathematicians studying combinatorial game theory often write out game trees, which illustrate the steps the game can follow, and use them to find optimal strategies.

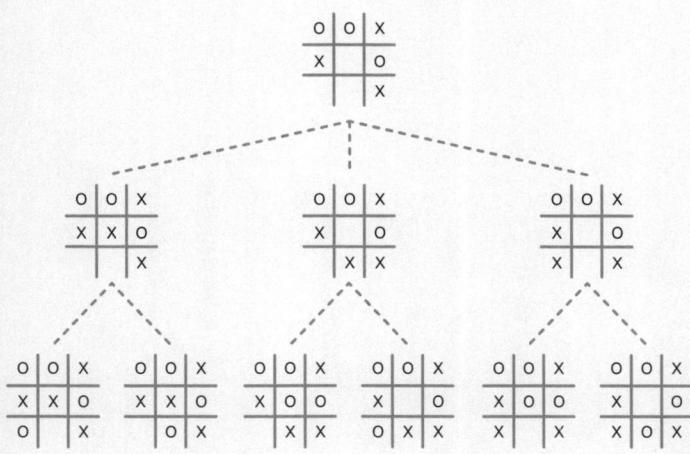

Part of a Noughts and Crosses game tree

It's also interesting to compare the sizes of different games. For example, there are 255,168 possible games of Noughts and Crosses, but Chess is a much more complicated game – even after each player has made only ten moves, there are already over 69,352,859,712,417 possible states the board could be in.

THE ART OF DECEPTION

HOW CAN YOU CHEAT AT CARDS?

Casino games are often based on probability and chance – from the spin of a roulette wheel to a shuffled deck of cards. But, while chance plays a major role, mathematics can tip the odds slightly in your favour.

THE GAME OF BLACKJACK

One of the most popular casino games is Blackjack. Each player is dealt two cards, with the goal of getting as close to a total of 21 as possible – each card is worth a set number of points (see opposite) and you add these points together to get your total.

You can then add cards to your hand one at a time – as many as you want, to try to get closer to a goal of 21 – but if your total goes over that, you're 'bust' and you lose that round.

Blackjack is played against a dealer, who has their own hand to maintain, and in a casino, the players bet money for each hand. If the dealer wins, the casino gets to keep everyone's money.

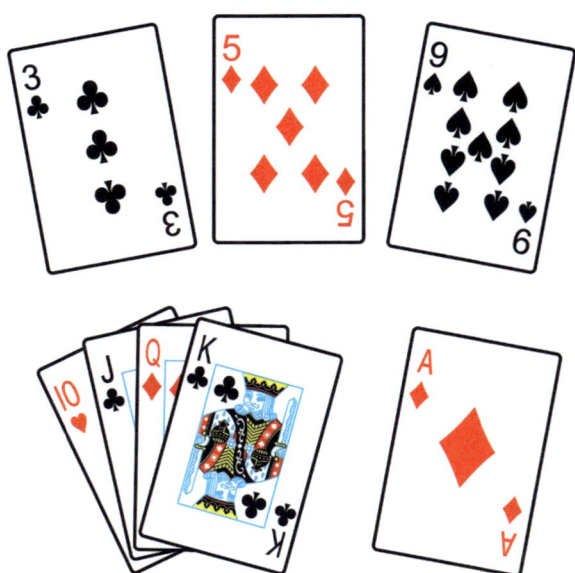

Card values

- Number cards (2–10): face value – so, a 6 is worth 6 points, and so on
- Picture cards (Jack, Queen, King): 10 points
- Ace: 11 points, or 1 point (if 11 would make you bust)

There's only one way to make 21 with two cards: an ace, and another card worth 10 (a ten or a picture card).

Example hand: stick or bust?

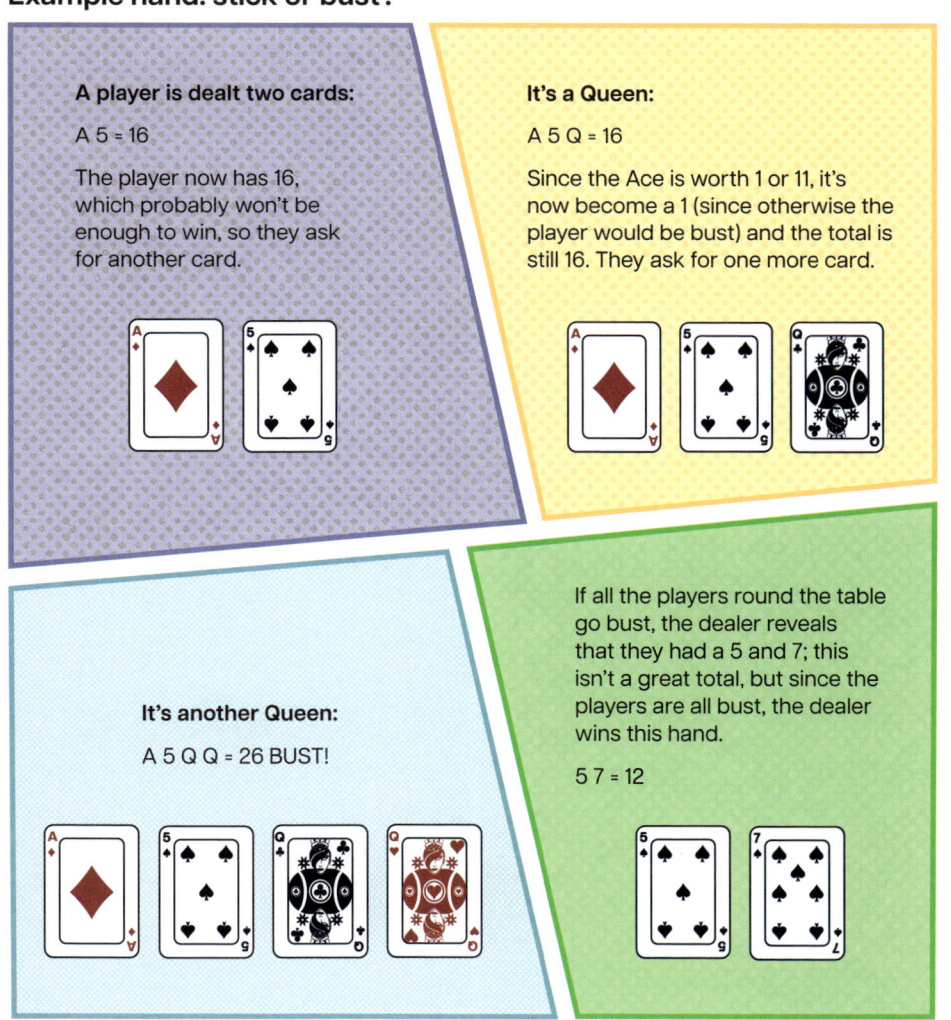

A player is dealt two cards:

A 5 = 16

The player now has 16, which probably won't be enough to win, so they ask for another card.

It's a Queen:

A 5 Q = 16

Since the Ace is worth 1 or 11, it's now become a 1 (since otherwise the player would be bust) and the total is still 16. They ask for one more card.

It's another Queen:

A 5 Q Q = 26 BUST!

If all the players round the table go bust, the dealer reveals that they had a 5 and 7; this isn't a great total, but since the players are all bust, the dealer wins this hand.

5 7 = 12

If the player chooses to 'stick' with the cards they have, it's the dealer's turn to choose. The dealer must take cards until their score is at least 17, then stick. This means they might go bust – if they're on 16, they still have to take more cards, and there's a chance that'll put them over 21.

If there's more than one player at the table, each one gets a turn to take as many cards as they want, or to stick. But even though the dealer has to play in a specific way, the casino has a built-in advantage:

- In the event of a tie, the dealer wins by default. This means that if all the players bust, even if the dealer also busts, they still win.

As with all casino games, the rules are in favour of the house, so on average over time the casino always makes a profit.

THE MATHS BEHIND BLACKJACK

Here we've broken down the distribution of card scores in a standard 52-card deck:

Score	Number of cards	Score	Number of cards
2	4 (2♣, 2◇, 2♡, 2♠)	7	4 (7♣, 7◇, 7♡, 7♠)
3	4 (3♣, 3◇, 3♡, 3♠)	8	4 (8♣, 8◇, 8♡, 8♠)
4	4 (4♣, 4◇, 4♡, 4♠)	9	4 (9♣, 9◇, 9♡, 9♠)
5	4 (5♣, 5◇, 5♡, 5♠)	10	16 (10 J Q K♣, 10 J Q K◇, 10 J Q K♡, 10 J Q K♠)
6	4 (6♣, 6◇, 6♡, 6♠)	11 (or 1)	4 (A♣, A◇, A♡, A♠)

While it looks like 10 is the most likely score, it'll come up only $16/52 \approx 30\%$ of the time. If you know how far away from bust you are, you can work out what your chances are of getting the card you want – so you can choose whether or not to bet big.

Estimating the odds

In our example earlier, the player was initially holding cards worth 16, meaning another card worth 5 or less would get them closer to 21, and anything over 5 would make them bust. There are 18 cards left in the deck worth 5 or less (three 5s, four 4s, four 3s, four 2s and three Aces – not including the five and Ace they're already holding), so there's an 18/50 chance of getting a card they want – not great odds, but it's still just over a third of the cards. And after they draw the second card and get a Queen, their total is still 16 and there are now 18/49 good cards, which is slightly better.

Except these probabilities aren't quite right: at this point, the dealer's 5 and 7 have also already been dealt, and there will be other players' hands being dealt from the same deck, so those cards are also in play. When you get dealt an Ace, it takes away that card from the deck – and other players are then less likely to be dealt one on their turns.

This means there's a trick you could use to gain a slight advantage in this game: it's called **card counting**, and involves keeping track of which cards have been played and keeping a mental note, so you can judge more accurately if the card you need is likely to come up next or not.

Counting cards

Roughly, the system involves scoring cards from 2 to 6 as '+1', cards 7 to 9 as '0' and cards 10 to A as '−1'.

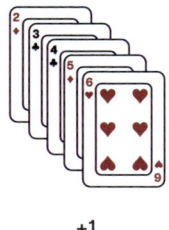      

+1   0   −1

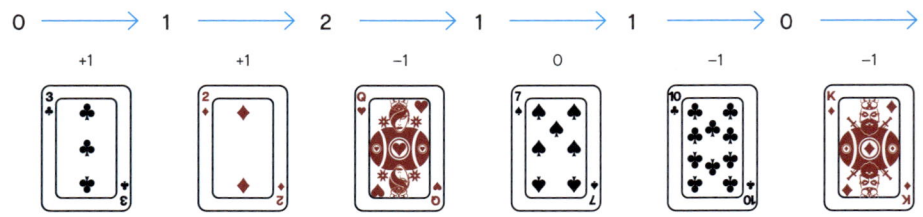

Card counting for the first ten cards of a game

The total starts at 0 for a newly shuffled deck, and then goes up and down as the cards come out, and gives the player a rough idea of what cards are left in the deck. A positive total means lots of smaller numbers have come out, and the deck is likely to have more 10s and picture cards; a negative total means the deck has a higher chance of giving you a low-numbered card next.

The figure above shows the first ten cards dealt in an example game. After these cards are dealt, the deck contains 42 cards, and five of the sixteen 10-point cards have been drawn, so the remaining cards on average have lower values – represented by the count being –3. So a player looking for a low number to make 21 would be more confident of getting it – and might increase their bet, or be more likely to ask for extra cards.

The advantage you gain from doing this is very slight – only a couple of percentage points of difference overall – but using this strategy could mean that over a long session you do slightly better than you would without it. It doesn't increase your chances of getting the cards you want, but it does mean that when your chances are better, you're aware of it and can play accordingly.

It's also very difficult! Concentrating on the total – while also thinking about your cards and what to bet, and trying not to make it obvious you're doing this – takes a lot of focus. Some people have been known to use technology to log data, like a computer hidden inside their shoe – they can press down or lift up their toes to log which cards have been played, and the computer activates a buzzer on a particular part of the foot to indicate what the current 'count' is, so they can use it to decide on their play.

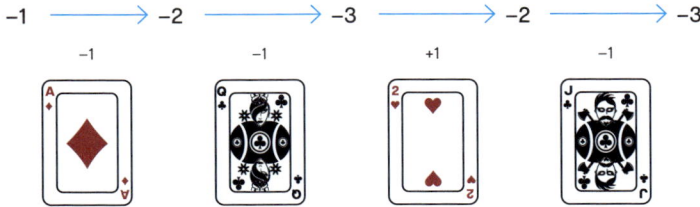

However, regardless of whether it's easy or effective, or whether you consider it 'cheating', card counting is definitely unfair on some people: as well as putting the casino at a disadvantage, you're also possibly cheating any other players out of a fair game (or, a game that's as fair as the casino wants to make it...)

Casinos take a dim view of card counting – although they're quite happy to manipulate the games in their own favour. While it's not technically illegal to count cards – although external devices like shoe computers are banned in some places – players suspected or found to be card counting can be expelled or banned from the casino, or even have their chips confiscated.

To prevent card counting, many of the more high-tech casinos now use a continuous deck: instead of a single deck of 52 cards that gets shuffled when it runs out, the dealer uses an automatic shuffling machine, with a constant supply of randomly shuffled cards from multiple decks shuffled together. Or casinos might instruct the dealer to shuffle the cards at certain points in the game, like when a player increases their bet.

These measures effectively eliminate the advantage gained by card counting – so even if you can 'cheat' at cards by card counting, it probably won't help!

HOW TO CATCH A CRIMINAL

One of the big questions in any criminal investigation is 'Where?' – more specifically, the question might be: 'where is the suspect?' or 'where is the missing person?' These sorts of questions often end up getting mathematical, falling loosely under the topic of **geolocation**.

CIRCLES AND ELLIPSES

Here are a couple of practical experiments you can try, which we'll relate back to our criminal investigation question.

1 Make a fixed loop of string or thread (around 10cm long when pulled tight is good, but don't use a rubber band – you don't want this to be stretchy).

2 Loop it around a fixed object on a flat surface (like drawing pin in a pinboard).

3 Use a pen or your finger to pull the loop tight and move it around, while keeping it tight.

4 Voila: you have traced out a perfect circle.

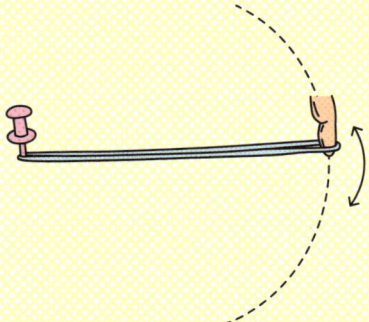

This probably isn't a surprise. One definition of a circle is 'the set of points at a constant distance from a fixed point', and that is exactly what our experiment has created. Now, try this:

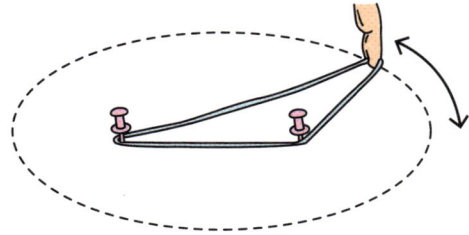

5 Grab the loop of string again.

6 Loop it around two fixed objects, some distance apart (like two drawing pins on a pinboard).

7 Pull the loop tight with a pen and move it around, keeping the loop tight around both the fixed points.

8 Voila, you have drawn something else: not a circle.

This time you have created an **ellipse**. Each fixed point is called a **focus**, or **focal point** (plural: foci). The string loop ensures that the sum of the distances from the two fixed points (foci) to the line you're drawing stays the same. The loop is the same size all the time, and so is the distance between the two fixed points – so the two remaining sides of the triangle you make with the string must add to the same sum all the time. This is the definition of an ellipse: 'the set of points where the *sum* of the distances to two foci is constant'.

TRILATERATION AND CIRCLES

Imagine you pick up a radio signal from a known criminal vehicle. It is usually impossible to know which direction you receive a radio transmission from – it just arrives. However, the signal is likely to contain a **timestamp** – an accurate record of the time it was sent.

If you also accurately know the time you *received* it, you can compare this to the timestamp.

The difference between the two times is the time it has taken for the signal to reach you, and multiplying this by the speed radio waves move at (the speed of light) will tell you the distance it's travelled.

This corresponds to our first experiment on page 56 with the loop of string and a single pin – if you know how far away it was, you can locate the source of the signal to be *somewhere on a circle, centred on your location*. If the same signal is also received at another location, you can generate another circle centred there. And since the signal was sent from a place on both these circles, then it must have been sent from one of the points where the circles overlap.

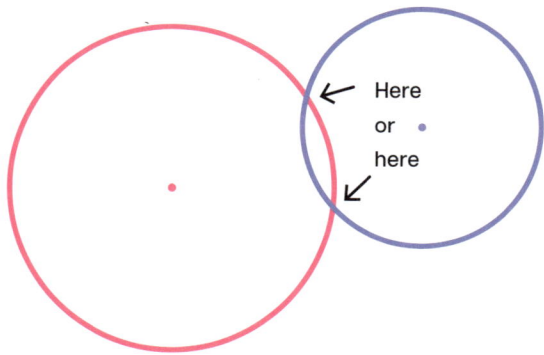

There are at least two big problems with this: you need to rely on the vehicle producing a signal with a timestamp, and you need to pick up the signal multiple times (at least twice, and maybe even three times to pin it down). So much for circles!

This process is called **trilateration**, because you need three length measurements to locate something (compare this word with 'triangulation', which uses three angles). It's mathematically elegant, but it also depends on the target being cooperative. This means it's great for tracking public aircraft, but less good if your target is a criminal on the run, or a combat target, intent on staying hidden.

SILENT SURVEILLANCE: ECHOES AND ELLIPSES

Let's consider a slightly different idea: imagine you send out a blanket signal from a transmitter, which happens to 'bounce' off your target at an unknown location. This may sound fanciful, but if the target contains a **transponder** – short for transmitter–responder – then that is precisely what it does. All aircraft have a transponder, which automatically sends out a response on receipt of a standard message.

Even if there's not a transponder handy, signals do also just bounce off things. Echoes from sounds will be familiar to anyone who's explored a cave (with lots of nice bouncy walls to reflect the sound back), and RADAR and SONAR (using radio and sound waves respectively) work on exactly this principle.

If the 'bounced' signal is then picked up by a receiver at another location, at a known time, you have some new information: the time it has taken (and hence the distance) from the original transmission point to the target, and then to the receiver. This corresponds to our second experiment with two pins – you know the sum of the two distances, so you now know the target is *somewhere on a particular ellipse with focal points at your transmitter and your receiver*.

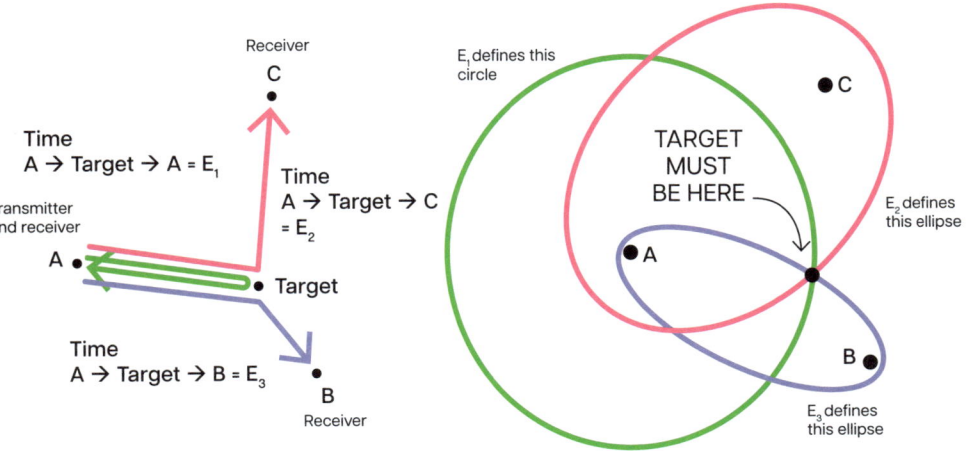

In the diagram on the left, a signal has been sent from point A, bounced off the target, and received at points A, B and C. The times between sending and receiving are used to generate the circle and ellipses in the diagram on the right, which overlap in only one place – the location of the target.

You still need to receive the bounced signal at three locations, but one of them could be your original transmitter (if it can also receive signals) – giving a circle around your own location, and you now need only to bounce the signal off your target, or exploit a transponder. This sort of thing is the idea behind echolocation (as observed in bats) and is sometimes called **elliptical trilateration**.

The glaring problem here is that you have to transmit a signal, or noise, at your target – and in doing so you may well reveal that you are on to them, which might not be desirable...

HYPERBOLIC HUNCHES

Finally, let's consider what would happen if you have a set of listening posts (receivers), which don't send out any signals at all, keeping your surveillance nicely discreet. If multiple receivers happen to pick up the same transmission (or noise), even if it doesn't have a timestamp built in, you are still getting some information.

Assuming you accurately note the time of arrival at each receiver, you can deduce the *difference* in time taken to get to each receiver, and therefore the *difference* in the distances to the two receivers from the target (as opposed to the sum, in the ellipse example).

If the difference in the times of arrival is precisely zero, then the target must be at a point equidistant between the two receivers (somewhere on the perpendicular line between the two).

If the difference is not zero, then you know that the target is closer to one receiver than the other. There's another mathematical curve, the **hyperbola**, which is the set of points where the *difference* of the distances to two foci is constant. Three receivers are enough to get at least two differences in time of arrivals, and hence (at least) two hyperbolae (there will be a third one available, because three receivers give three possible pairs: $^3C_2$ again). You should find the target at the point where they intersect – unlike with the circles, there is only one point of intersection.

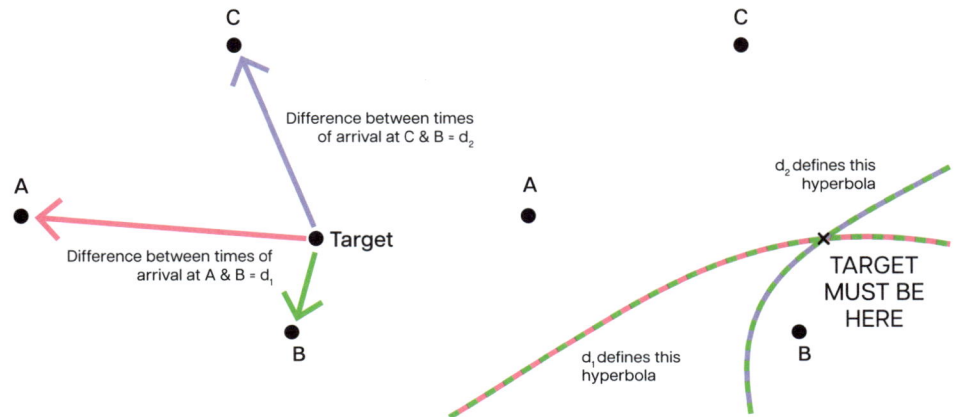

Difference between times of arrival at C & B = d_2

A

Difference between times of arrival at A & B = d_1

● Target

B

C

A

d_2 defines this hyperbola

X
TARGET MUST BE HERE

B

d_1 defines this hyperbola

In the diagram on the left, a signal has been picked up by receivers at A, B and C. The difference between the arrival times at A and B generates one hyperbola, and the difference between the arrival times at C and B generates another.

The curves we've seen here are all cross-sections of a cone – called conic sections, they have been studied since the times of Ancient Greek mathematics.

These ideas, along with various upgrades to make them work in three dimensions, are used in global positioning systems (GPS) as a matter of course. They are not just used for finding fugitives or catching criminals – every time you check your location on your phone it does some serious mathematics: with conic sections in the background.

circle
ellipse
parabola
hyperbola

CAN SOMEONE TELL IF YOU'RE FIDDLING YOUR TAXES?

Maths is a great tool for when you're handling money, or working out your finances. One situation where you'd definitely get your calculator out is if you're completing accounts for a tax return – whether for an individual or a business. This involves a lot of adding up and manipulating numbers, in order to report sales and expenses for the year.

All these numbers, totals and calculations have to be correct, which is why businesses are required to keep receipts and detailed records of all their transactions. But if you can't find the right papers, or if you're the unscrupulous sort who wants to avoid paying the right amount of tax, you might be tempted to change the numbers, or make up your own plausible-sounding numbers to fill in the gaps.

You might think there's no way for an auditing organization to detect this kind of fraud, but it turns out there is, and it's thanks to a piece of mathematics called **Benford's law**.

BENFORD'S LAW

Benford's law is a property of data distributions that applies to many different types of data – from house numbers to amounts of money. Given a set of data – like a list of expenses claims – it turns out to be very useful to analyse specifically the *first digits* of each number, sometimes called the **leading digits**.

If we picked a set of random numbers between 1 and 99, you might expect that the leading digits of the numbers would also appear random. And you'd be right – each has a roughly equal chance of being any one of the digits 1 to 9, and each one would show up about 1/9 of the time.

But for data like expenses from real accounts, this isn't what we observe. Here's a graph showing how the first digits of each number are distributed in a real dataset:

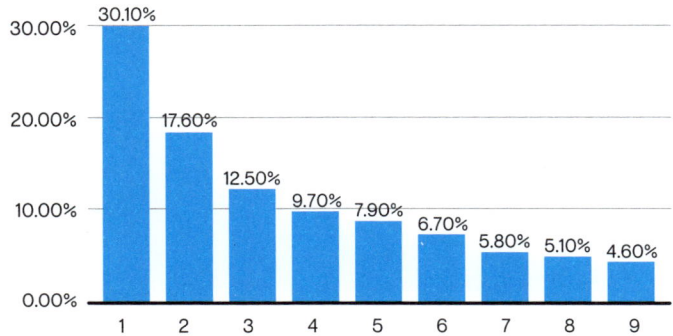

It turns out that in certain types of real data, numbers have the first digit 1 around 1/3 of the time (30.1%); their first digit is 2 around 17% of the time, and so on. There's a gradual decline until numbers with the first digit 9, which show up less than 5% of the time.

This seems surprising – just because the numbers happen to be expenses doesn't mean they're not random numbers. And this isn't because expenses happen to be skewed in a particular way – it's a feature of numbers that represent a quantity. The big clue is that our original example had a very specific limit on how high the numbers could be (a maximum of 99). In the messy real world of expenses, there's no such limit, and the numbers tend to come from a much larger range of options, over several orders of magnitude.

This same distribution of leading digits shows up in a range of different types of data. If you picked a list of random places in the world (of any size – country, city or town) and look at the sizes of their populations, you'd see the same distribution of first digits. It also works for the heights of the world's tallest buildings, and the personal-best times in seconds for attempts at Rubik's cube solving at World Cube Association events. It even happens if you look at the front page of a newspaper and write down all the numbers you can see on it – since these numbers are all taken from different sets of unknown size, they'll be distributed in the same way.

This won't work if your data is from a set with tight constraints on it – for example, the heights of people in centimetres. These all usually sit within a particular range – from about 1 metre to 2 metres – so you don't see the same distribution of first digits. The best examples are numbers that are taken randomly from a large range of possibilities.

WHY DO LEADING 1s APPEAR SO OFTEN?

One way to understand this is to think about house numbers. If a street has nine houses on it, numbered 1 to 9, we'd expect a flat 1/9 distribution.

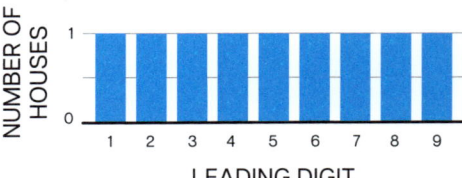

But if we add another house, so the street goes from 1 to 10, there are now two houses with first digit 1:

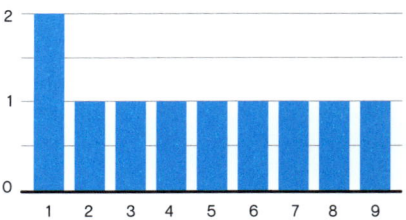

If we keep adding houses, the next nine houses all start with 1, so we'd get to a point, at 19 houses, where most of them (over half) start with 1.

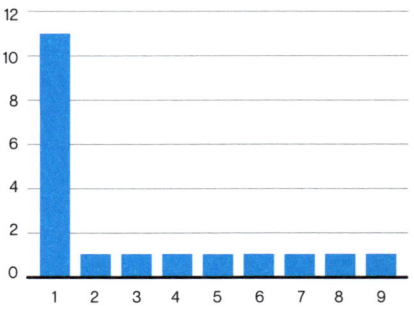

But from there, all the way up to 99, there are no more houses that start with 1, so adding more houses will reduce this proportion again (and this is the graph we'd see for our earlier example of random numbers up to 99).

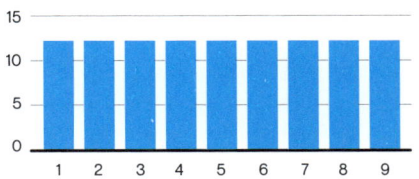

If we plot the proportion of houses that start with 1 against the length of the street, we see a sawtooth-like shape – the proportion varies between 1/2, or 50% (when we are just about to get to numbers that start with a 2, like the 20s or the 200s) and 1/9 (11%).

And if the data we're looking at is from the real world, it's like picking a house number at random from a street of unknown length: expenses can take many different values, but the smallest they can be is 0, and the largest depends entirely on what type of expense it is. This means the number of leading 1s we'd expect to see across a large dataset is the average of the values on this graph – the horizontal line that comes in about 1/3 (33%).

If we do a similar analysis for numbers that start in 2, 3 and so on we get averages that correspond to the distribution given by Benford's law.

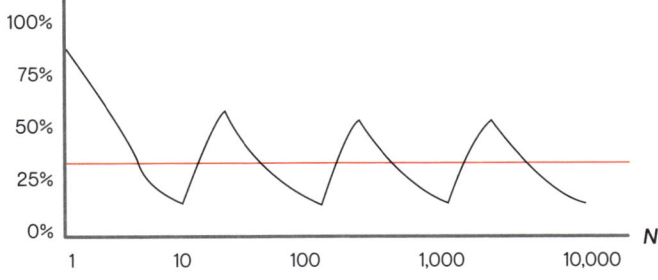

Proportion of numbers up to N that begin with a 1

CAN YOU FAKE IT?

All this means that if you send in your tax return with made-up numbers, it's likely they won't match this distribution – if they've been picked randomly, they will have random first digits, which won't obey a Benford distribution. Above is a plot of the distribution of first digits of a real expenses report for one year, for one unspecified self-employed mathematician/author.

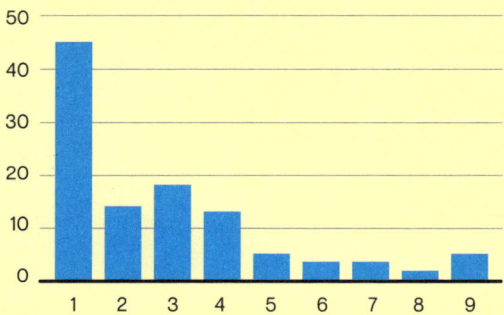

This distribution is very similar to the one given by Benford's law, and in general real data for quantities that cover several orders of magnitude will always display this kind of distribution. Forensic accountants use rules like Benford's law, and other similar mathematical tricks, to spot when people have been telling numerical fibs!

SO...

...can someone tell if you're fiddling your taxes?
Probably, yes.

ARE THESE STATISTICS LYING?

Here are four sets of data. Each set has two variables, x and y, which may or may not have some relationship between them.

Do you notice anything?

Dataset I		Dataset II		Dataset III		Dataset IV	
x	y	x	y	x	y	x	y
10	8.04	10	9.14	10	7.46	8	6.85
8	6.95	8	8.14	8	6.77	8	5.76
13	7.58	13	8.74	13	12.74	8	7.71
9	8.81	9	8.77	9	7.11	8	8.84
11	8.33	11	9.26	11	7.81	8	8.47
14	9.96	14	8.1	14	8.84	8	7.04
6	7.24	6	6.13	6	6.08	8	5.25
4	4.26	4	3.1	4	5.39	19	12.5
12	10.84	12	9.13	12	8.15	8	5.56
7	4.82	7	7.26	7	6.42	8	7.91
5	5.68	5	4.74	5	5.73	8	6.89

You might have noticed that three of the datasets have exactly the same numbers in the 'x' columns, but beyond that it's hard to spot anything meaningful. Raw tables of numbers are not helpful for seeing trends in data – but there are some statistical tools that can help us out.

SUMMARY STATISTICS

To understand or compare datasets, statisticians use summary statistics – numbers that describe features of the data without listing every point.

Measures of central tendency (averages)

- The **arithmetic mean**: sum all the data then divide by n, where n is the number of pieces of data. This is what most people think of when they use the word average, or mean. The AVERAGE() command in spreadsheets does exactly this – which is misleading, because there are many other 'averages' and 'means'.

- The **geometric mean**: multiply all the data together, then take the nth root, where n is the number of pieces of data. This is useful for averaging things that work together by multiplication, rather than addition (such as interest rates in financial calculations).

- The **mode**: the most commonly occurring piece of data.

- The **median**: the piece of data that's in the middle of the set, if you put it in size order.

Measures of spread (how scattered the data is)

- The **range**: how far apart the maximum and minimum data points are.

- The **interquartile range**: how far apart the points 1/4 and 3/4 through the data are. This is sometimes more helpful than the range, because it dodges extreme outliers, which can make the range unrepresentative.

- The **standard deviation**: roughly, a measure of how far away the data is from the mean. The smaller it is, the more values will be clustered close to the mean.

Correlation and connections

- The **correlation coefficient**: a measure, between –1 and 1, of how well correlated two datasets are, with 1 being perfect positive correlation (when one goes up, so does the other), –1 being perfect negative correlation (when one goes up, the other goes down), and 0 being no discernible correlation.

- A **regression line**: or the 'line of best fit', which uses a mathematical technique to suggest a possible equation for how the datasets relate to each other, if they're well correlated.

With these tools we can summarize the datasets we saw earlier. Using summary statistics should give us an idea of what a dataset is telling us, without having to look at each individual data point, and make it easier to compare different datasets.

Here are the arithmetic mean, the standard deviation, the correlation coefficient and the equation of the linear regression line for each dataset.

Do you notice anything this time?

	Dataset I		Dataset II		Dataset III		Dataset IV	
	x	y	x	y	x	y	x	y
Mean	9.00	7.50	9.00	7.50	9.00	7.50	9.00	7.50
Standard deviation	3.32	2.03	3.32	2.03	3.32	2.03	3.32	2.03
Correlation coefficient	0.82		0.82		0.82		0.82	
Regression line	$y = 0.50x + 3.00$		$y = 0.50x + 3.00$		$y = 0.50x + 3.00$		$y = 0.50x + 3.00$	

Each dataset has the same mean, for both x and y. Each one has the same standard deviation, for x and y. And the same correlation coefficient. And the same regression line.

So...are the datasets exactly the same?

Of course not! They are clearly not the same. This was exactly what Francis Anscombe, who constructed the datasets in 1973, wanted to emphasize: *summary statistics are not all you need to know.*

Anscombe advocated looking at plots of the data to get a better picture. Let's follow his advice. We can plot each dataset on a graph, with one variable on each axis.

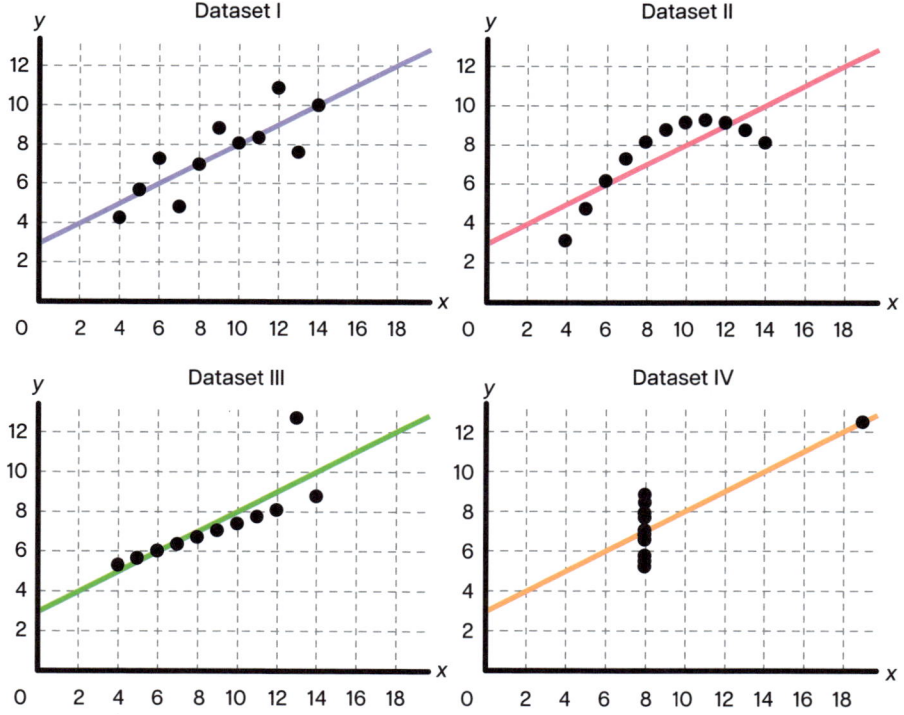

With these pictures you can better appreciate what's going on.

- Dataset I feels like a collection of loosely correlated data points.

- Dataset II is clearly not a linear relationship, but there's some connection between the two sets – the shape of the data looks like a parabola.

- Dataset III appears to be a linear relationship with one obvious outlier (a mistake?).

- Dataset IV has one variable that doesn't vary at all, indicating no relationship between the variables, but again with one extreme outlier.

Even though these datasets look very different, the summary statistics are the same in each case. It's even arguable that some of these statistical techniques should never have been used. If you looked at the pictures first, it would seem odd to suggest finding a straight line of best fit for some of them, particularly sets II and IV.

The fascinating thing about Anscombe's now-famous quartet of datasets is that no-one is sure exactly how he created them. It remains a useful teaching example, and Anscombe deserves credit for doing the manual hard work required to create such an example.

Since then, with the application of more modern coding techniques, several people have pushed the idea further.

- The Datasaurus Dozen – if you graph an initial set of data by Alberto Cairo it looks like a dinosaur (the Datasaurus, on the left of the picture). Researchers

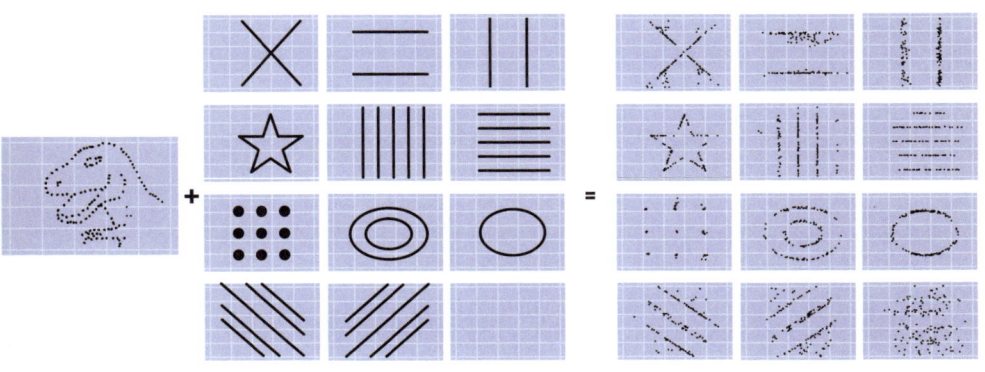

Justin Matejka and George Fitzmaurice created twelve new datasets *with exactly the same summary statistics*, each one forming a different shape.

- The Databet – inspired by the Datasaurus Dozen, Matthew Scroggs created an entire English alphabet where each letter is the graph of a set of data points, all of which have the same summary statistics.

HE EVEN MADE IT INTO A FONT

While this is fun, the mathematical techniques used to create artificially interesting datasets like these – called **simulated annealing** – involve repeatedly moving the data points by tiny amounts, keeping the statistics the same, and are widely used in machine learning, as a way to find approximately optimal solutions to otherwise intractable problems.

In conclusion:

ARE THE SUMMARY STATISTICS LYING? No. They're summarizing the data – but they don't always tell the whole story.

ARE THE STATISTICIANS LYING? Probably not, since the major feature of a statistician's job is to be aware of this sort of problem, and do better at representing the important features of data.

ARE OTHER PEOPLE LYING WITH STATISTICS? Yes. Maybe because they have misunderstood something, or maybe because they want to tell a story a certain way. This business is as old as humanity, and the only way to avoid it is to make sure you understand the statistical tools you're using. In this case: draw a picture.

CAN MATHS HELP YOU KEEP A SECRET?

The world is full of information, and some of that information is in the form of secrets. Thankfully, mathematics gives us some tools to be able to communicate certain pieces of information without revealing others. And, surprisingly, this doesn't always require complicated encryption algorithms or fancy computing.

Let's start with a simple but clever trick.

SECRET SALARIES

Imagine a group of people who all have similar jobs, and they want to find out what the average salary for people who do that job is – but without divulging their individual salaries to any one person.

You might think this is impossible, but here's a method.

- First person picks a random large number, adds it to their salary and writes it down.

- They pass this to the next person, who adds their salary and passes the result on.

- This repeats around the whole group until everyone has added their salary; then someone passes it back to the first person.

- If that person subtracts their original random number, the result will be the sum of everyone's salaries; dividing this by the number of people will give the average.

At no point in this process can anyone find out someone else's salary, since they only see a number that's a sum including a random quantity they can't know; the first person doesn't see the intermediate totals, so they can't use their knowledge of the random number to find out anyone's salary, and since each person only receives and sends a single number, they can't compare any to work out anything in between.

I CHOOSE THE RANDOM NUMBER £15,500, AND ADD MY SALARY WHICH IS £30,100, SO MY TOTAL IS £45,600.

MY SALARY IS £28,500 SO I'LL ADD THAT ON TO GET £74,100.

MY SALARY IS £35,000 SO I'LL ADD THAT ON TO GET £109,100.

I'LL SUBTRACT MY ORIGINAL NUMBER, WHICH WAS £15,500, SO WHAT'S LEFT WILL BE THE SUM OF EVERYONE'S SALARIES.

This allows us to communicate one specific piece of information (the average) without any unwanted information changing hands. There's a general category of mathematical ideas that allow us to pass on facts, or prove things, without giving away unnecessary information; they're sometimes called **zero-knowledge protocols**.

WHAT ARE ZERO-KNOWLEDGE PROTOCOLS?

A zero-knowledge protocol is a way for one person (the 'prover') to convince another person (the 'verifier') that a given statement is true, without giving away any other information.

This statement could even be something like 'I know the answer to this question', where the prover can prove they know the answer without giving away what it is.

Imagine a children's 'look and find' book, where each page has a large and detailed drawing of a scene full of characters, and the puzzle is to find a particular object or character on the page. If you want to prove you've solved one of these puzzles, but not spoil the answer, all you need is a large sheet of paper – more than double the dimensions of the book – with a small hole in it.

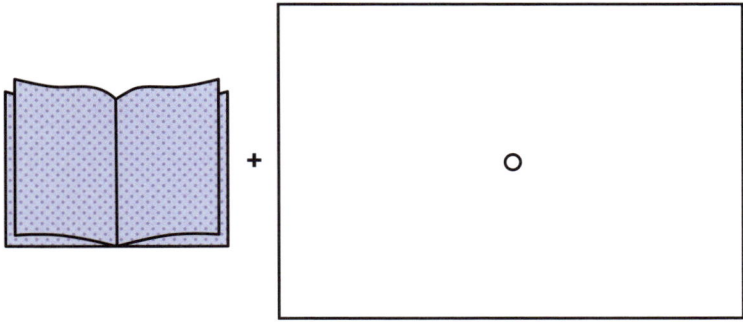

You can hide the entire book (page) under this large sheet of paper, but then carefully reveal the desired character through the small hole. You've proved that you know the character's location, but the paper is concealing the relative location of the book underneath, so you don't give away where on the page the character is.

SECRET SUDOKU

A more complex zero-knowledge proof protocol can be used to prove a statement like 'I know the solution to this sudoku'. A sudoku puzzle is a 9-by-9 grid in which every row, column and 3-by-3 subgrid needs to have the numbers 1 to 9 in once each.

An unsolved sudoku will have some numbers – 'clues' – already written in some, but not all, of the cells. Imagine a prover, Prue, who knows how to solve the sudoku and wants to show this to a verifier, Vera, but without the verifier finding out how to solve the sudoku.

Prue prints the sudoku on a piece of paper and writes the solution using a different colour pen (grey in this case) – so any square with a clue in it will have that number written in the original colour, and the rest of the squares will have a number in the new colour.

Vera can then request 'rows', 'columns' or 'subgrids'. If she chooses 'rows', Prue grabs a pair of scissors and cuts the grid up into separate rows, then snips each row into individual squares and puts them in an envelope. Vera knows where the clues were (so for each row, she can check that the numbers given as clues for that row are in the original colour) and can also check that each envelope contains the numbers 1 to 9 once each.

If Vera had picked 'columns' or 'subgrids' instead, Prue could just cut it up in a different way – but any one of these can be checked, and if there's a valid solution you'd get the same result.

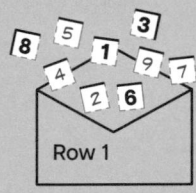

Row 1

SECRET SANTA

You may also have encountered this sort of problem when organizing something like a Secret Santa gift exchange. The gift givers need to remain secret, despite information about who each person has to buy for needing to be exchanged. There are various ways to implement this, but mostly people put all the names of the participants on bits of paper in a bag, and everyone draws one out. This will give you a random assignment, but it does have some issues. For example, there's a chance you might have to buy a gift for yourself!

A fun physical way to get around this problem is to use a stack of cards numbered from 1 up to the number of people, with the number written on both the top and bottom half of the card (and a non-symmetrical marking on the back, so you know which way round it is when face down).

Start by shuffling the cards to put them in a random order, then deal them all out in a line face down, in the same orientation, and stick the cards together using tape. Then cut each card in half (sticking together the two half-cards from the ends to make a final pair, without looking at them).

Now pass a half-card-pair to each person. This will have a number in the top half, which tells each person what their own number is. The person they're buying for is the number on the bottom half, which only they know. Everyone announces their own number so people can find out who they're buying for. This gives information only to the person who's buying for you, and doesn't let anyone find out who is buying for them.

In this case, the assignment of Santas to recipients is a cycle: each person is buying for the next person around in the cycle, so you can guarantee nobody is buying for themselves, but the assignment is still random (based on the shuffle you gave the cards to start with). Mathematics helps you keep a secret, and saves Christmas again!

THIS STATEMENT IS A LIE

Imagine you meet two people: Ali and Beth. Ali says to you 'Beth is lying'. Beth says 'Both Ali and I are telling the truth'. Which one is telling the truth?

This puzzle uses logical principles to create a confusing situation: the truth of particular statements can't be known immediately, especially if the statements themselves are about whether or not someone is telling the truth. Some statements are even so unclear they create mathematical paradoxes! We often need to use a little logic to work out what's really going on.

In this case, there are four possible options, assuming each statement is either true or false:

1. Ali and Beth are both telling the truth.	2. Ali is lying, and Beth is telling the truth.
3. Ali is telling the truth, and Beth is lying.	4. Ali and Beth are both lying.

It can't be the case that both of them are telling the truth, since the statements they make are contradictory – one says Beth is telling the truth and one says Beth is lying, and these can't both be true. So we can rule out statement 1.

Statement 2 is also impossible, since if Beth is telling the truth, Ali must be telling the truth, and we've assumed Ali is lying. So that's ruled out too.

By a similar logic we can rule out the statement 4, since if Ali is lying, the statement 'Beth is lying' must be false, and that contradicts our assumption that Beth is also lying.

By this process of elimination, we can see that statement 3 is the only one that can possibly be consistent – Ali states that Beth is a liar, and then Beth says something that would therefore be a lie.

KNIGHTS AND KNAVES

This kind of logic puzzle, often called a 'Knights and Knaves' puzzle, has been around for a long time, but was popularized in puzzle books in the late 1970s. They often involve a situation where everyone is either a full-time liar (a knave), or a 100% truth-teller (a knight), since this simplifies the logic and rules out the possibility that someone might lie only some of the time (which in real life is often the case – but we're working in the mathematical world of logic here).

In each case, you can consider the four possible options – two truth-tellers, a truth-teller and a liar, a liar and a truth-teller, or a pair of liars – and rule out situations that are inconsistent with the statements until you work out what must be happening.

KNIGHTS AND KNAVES

An island is inhabited by two kinds of people: truth-tellers and liars. Truth-tellers always tell the truth, and liars always lie.

Puzzle 1: Sam and Pam
You meet two inhabitants: Sam and Pam. Sam claims, 'Pam and I are not the same kind of person.' Pam says, 'Out of me and Sam, exactly one of us is a truth-teller.' Who is telling the truth?

Puzzle 2: Billy and Tilly
Later, you meet two other inhabitants: Billy and Tilly. Billy tells you, 'Either Tilly is a truth-teller or I am a liar.' Tilly tells you that Billy is a liar. Who is telling the truth?

Puzzle 3: Ali and Sally
Even later, you meet two more inhabitants: Ali and Sally. Ali tells you, 'At least one of these things is true: that I am a truth-teller or that Sally is a liar.' Sally says, 'I would say that Ali is a liar.' Who is telling the truth?

LIAR PARADOX

There's one statement you won't ever find in a logic puzzle like this: you'll never see anyone make a statement along the lines of 'I am a liar' (in the specific situation where that means every statement they make is a lie).

This is because the logical consequences of this statement don't make sense. If the person is telling the truth, then they must be lying – and if they're lying, then they must be telling the truth...there's no consistent way to resolve it, and the result is a paradox. Often called the 'liar paradox', it can be most simply stated as:

'This sentence is false.'

Mathematicians working with logic need to avoid these kinds of paradoxes, since there's no way to make sense of the situation in that case. There are some fun ways to use this kind of paradox to make impossible, paradoxical questions. For example:

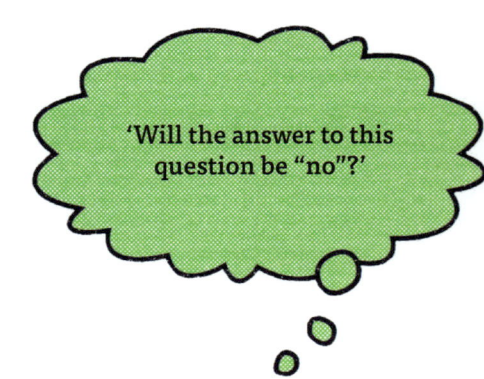

'Will the answer to this question be "no"?'

This one's impossible to answer, since if the answer is 'yes', that means the answer must be 'no', and vice versa. Another step on from this allows us to create an even more torturous paradoxical question:

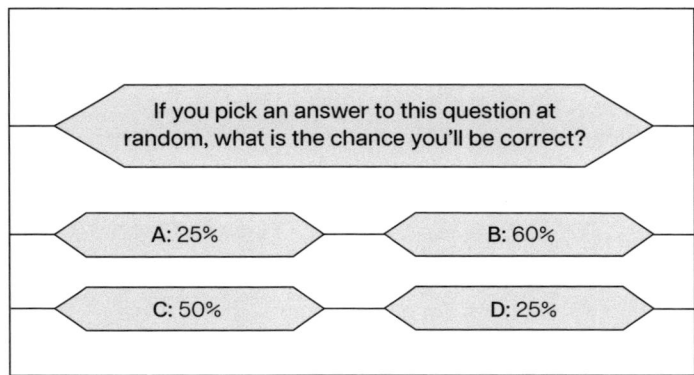

If you pick an answer to this question at random, what is the chance you'll be correct?

A: 25%   B: 60%

C: 50%   D: 25%

In this question, there are four possible options, and picking one at random would mean you have an equal chance of picking each one. So, if there were four different answers, one of which is correct, you'd have a 25% chance of choosing correctly. But this question isn't quite so simple!

The question has been carefully set up so that none of the possible answers give a consistent probability: picking A or D would give an answer of 25%, but since two of the answers give this same value, this will happen with a 50% chance overall, so it's wrong; picking C would happen with a 25% chance, as would B, but neither of these are correct either.

While it seems like a simple enough question, it turns out the answer is 0%, since none of these answers will work!

MONEY, MONEY, MONEY

CAN YOU RETIRE EARLY?

Let's test your financial intuition. What do you make of this story, told to us by our friend – author and mathematician Rob Eastaway?

THE POCKET MONEY DILEMMA

A girl asked her parents for more pocket money. They said no, and countered that she should start to earn her pocket money by doing chores.

Annoyed that her request had backfired, the girl offered her parents two options. She would do the chores, in return for a signed month-long contract, agreeing to either:

OPTION 1: £100 a week, paid each Sunday.

OPTION 2: 1p on the first day, 2p on the second day, and doubling each day until the end of the month.

The parents agreed to the more reasonable-sounding Option 2, and signed a contract to this effect.

Realistic or not, this is a salutary lesson in understanding growth.

Week 1: Nothing to worry about

After one week, the following payments had been made, with a total of £1.27 paid out.

Mon	Tues	Weds	Thurs	Fri	Sat	Sun
1 1p	2 2p	3 4p	4 8p	5 16p	6 32p	7 64p

Week 2: Still manageable

So far, so good. In the second week, the girl magnanimously rounded the amount down to £10 (rather than £10.24) on the 11th day. By the end of two weeks though, a total of £160.23 is beginning to feel quite generous.

Mon	Tues	Weds	Thurs	Fri	Sat	Sun
1 1p	2 2p	3 4p	4 8p	5 16p	6 32p	7 64p
8 £1.28	9 £2.56	10 £5.12	11 £10	12 £20	13 £40	14 £80

Week 3: Alarm bells are ringing

After three weeks, the family savings are probably taking quite a hit, with a total of over £20,000 (even rounding-down to £10,000 for simplicity on the 21st).

Mon	Tues	Weds	Thurs	Fri	Sat	Sun
1 1p	2 2p	3 4p	4 8p	5 16p	6 32p	7 64p
8 £1.28	9 £2.56	10 £5.12	11 £10	12 £20	13 £40	14 £80
15 £160	16 £320	17 £640	18 £1,280	19 £2,560	20 £5,120	21 £10,000

Week 4: Financial ruin

By day 25, they're in 're-mortgage the house' territory, and the terrible feeling of crossing the £1 million mark on the 28th is made worse by the awareness that Option 1 would have only cost them £400 by this point, and there are still three days to go.

Mon	Tues	Weds	Thurs	Fri	Sat	Sun
1 1p	2 2p	3 4p	4 8p	5 16p	6 32p	7 64p
8 £1.28	9 £2.56	10 £5.12	11 £10	12 £20	13 £40	14 £80
15 £160	16 £320	17 £640	18 £1,280	19 £2,560	20 £5,120	21 £10,000
22 £20k	23 £40k	24 £80k	25 £160k	26 £320k	27 £640k	28 £1.3 million
29 £2.5 million	30 £5 million	31 £10 million				

Many people – even those fully acquainted with what happens when you repeatedly double things – are surprised by how quickly this situation changes from feeling reasonable to utterly ridiculous.

Exponential growth is a term used to describe the situation where something grows according to a multiplying factor on a regular basis – for example, doubling each day. It's often used in modelling things that grow or decay over time, like populations, investments or radioactive decay. It is exactly what is meant by the phrase **compound interest**, where investments earn interest payments, and then these form part of the amount for the next interest calculation.

The cheeky daughter's request translates to an interest rate of 100% per day. this is a ridiculously high interest rate, as the parents *should* have known. If any part of you remains surprised about the outcome, consider checking your understanding before taking out a loan.

CAN YOU ACTUALLY RETIRE EARLY?

Let's shift from fiction to finance. If we want to retire early, what does the maths tell us?

Understanding compound interest

Let's work with an interest rate of 5% – this means, loosely, that if you borrow money as a loan (e.g. a mortgage, to buy a house) you'll have to pay back that amount, multiplied by 1.05 for every year you have that loan. The detail gets messy, because you'll almost certainly pay it off a bit at a time, but that's the basic principle.

Conversely, if you invest money (such as in a savings account) you might expect to see that amount increase by a multiplier of 1.05 each year. In practice, the savings rates are always slightly less than the loan rates – otherwise everyone would be borrowing money to then put it in a savings account.

But how does that help with early retirement?

If you can generate enough *passive income*, say, through savings or investment, you might not need to keep working. For example, If you had £800,000 invested, at a 5% interest rate, you would earn £40,000 in interest each year, before tax.

How can you generate that lump sum for when you want to retire? The advice is pretty simple.

Key takeaways for early retirement

1 **REDUCE YOUR COSTS:** This does two important things: it reduces your expectations for the amount you need to live off each year (so your target income can be lowered) *and* it helps with the next piece of advice...

2 **PUT MORE IN YOUR SAVINGS:** The earlier you can put money aside in savings, the more effect the compound interest will have on helping it grow.

Saving over time: an example

Let's say you put £5,000 in savings at the start of each year, and earn 5% interest at the end of each year, for 40 years. To work out the amount you'll have after 40 years, you can think about the final year's contribution first, and then work backwards:

$$5{,}000 \times 1.05 + 5{,}000 \times 1.05^2 + 5{,}000 \times 1.05^{\circled{3}} + \ldots + 5{,}000 \times 1.05^{39} + 5{,}000 \times 1.05^{\circled{40}}$$

This is the contribution from the final year's addition of £5,000

This is the contribution from 3 years ago, which has earned interest for 3 years

This is the contribution from 40 years ago, which has earned interest for 40 years

This long calculation can be simplified to:

$$£5{,}000 \times (1.05 + 1.05^2 + 1.05^3 + \ldots + 1.05^{39} + 1.05^{40})$$

The expression inside the bracket is a **geometric series** – a sum of things that increase by a multiplying factor each time – and there's a formula to calculate it in one step. Applying the formula gives:

$$£5{,}000 \times 1.05 \times \frac{(1.05^{40} - 1)}{(1.05 - 1)} = £5{,}000 \times 126.8$$

Your £5,000 a year has become 5,000 × 126.8, which is £634,198.80, over 40 years. If you then put this total in a savings account giving you 5% interest, it would pay you around £32,000 a year, before tax. Not too bad.

IMPORTANT CAVEATS

You may have to pay tax on your interest income each year (unless it's all in a tax-free savings account, like an ISA) which will reduce the amount you earn.

The calculation assumes that you are consistent with how much you save each year, and that the interest rate stays the same for 40 years.

Inflation means that whatever you've saved is likely to end up less valuable than it was when you first saved it (so even if £32,000 is a decent annual salary now, it might not be in 40 years).

THE EARLY RETIREMENT FORMULA

Assuming you've heard all the warnings and caveats, you can put your own numbers in the following equation to calculate what amount (S) you need to put aside each year to achieve your desired retirement income. That done, you'll need to stick to your plan for the next n years, and account for taxes and inflation along the way. Good luck!

$$S = \frac{T \times p}{(1+p) \times ((1+p)^n - 1)}$$

- S = amount you would need to put aside each year

- p = percentage interest you can get on your savings, as a decimal (e.g. 5% = 0.05)

- n = number of years you can do this (e.g. years to your proposed retirement)

- T = the total lump sum required (e.g. the £800,000 from earlier)

SO...

if you want to retire early, the advice is exactly what you might have already guessed: spend less and save more.

HOW TO FAIRLY SPLIT THE BILL

When going out for a meal with a large group, you'll often find yourself having to do a calculation at the end: deciding who owes what, so you can all pay your share of the bill.

FAIR DIVISION

If you're all in the mood for a little calculation, the fairest way is for everyone to work out how much their own food and drinks cost, and pay their own share. But if the group you're dining with wants to save time on calculations, they might agree to split the bill evenly – dividing the total spent by the number of people dining. Some restaurants now helpfully include this calculation on the bill for you.

While this does make the maths easier, it can lead to some unhelpful incentives for the diners. If everyone knows in advance the bill is going to be split equally, the temptation to order more drinks, or more expensive food, or extra sides, creeps in – you'll be splitting that cost between everyone, so if you order a more expensive meal than everyone else, you personally get more of a bargain.

This is obviously unfair on your table mates – they're paying more money and getting less food. Is there a way to split a bill that's less complicated than working out everyone's order separately, but fair enough to mean nobody's getting short-changed?

The Cheese Sandwiches and Lobster Restaurant

Cheese sandwich	£4
Cheese and onion sandwich	£4.50
Full lobster with sides	£30
Cheese and pickle sandwich	£4.22
Total	£42.72
Split between 4 diners:	£10.68 each

$$1 \times £5 = £5 \qquad 3 \times £5 = £15 \qquad 6 \times £5 = £30$$

THE 'UNITS' METHOD

One approach is to think about meals in 'units'. One main course might be considered to be two units, or three units if it's something fancy like a steak; a starter or side, one unit; a soft drink might be one unit, and a more expensive drink might count as two. Each person can count up how many units they ordered much more easily than calculating their own bill, and it gives a rough measure of who ate and drank the most.

If everyone makes an estimate of their meal in units, you can add up these numbers to get the total for the whole table. Then, once the bill arrives, divide the full cost by the number of units, to work out how much one 'unit' cost, and each person pays for the appropriate number of units. If a unit ends up being close to a nice round number like £5, the maths will be much quicker than fiddling with precise amounts.

But maybe this method feels too much like hard work – after all, going out with friends should be fun, and functional calculations like this aren't that interesting. So here's a more exciting way of working out who pays what, which means people can still order different amounts of food and drink, and not feel like anyone is paying more or less than they should.

Person 1 has a small salad: 1 unit; person 2 has a main and soft drink: 3 units; person 3 has an expensive main (3 units) and three sides: total 6 units. The total bill cost is £50 and there are 10 units in total. Therefore the cost of 1 unit = £50/10 = £5.

THE 'RANDOM ROUND' METHOD

When the bill arrives at the end of a meal, it'll have a list of what people ordered, and a total at the bottom. Everyone should be able to remember which of the items on the menu were consumed by which diner. Then find a random number generator (on a calculator, or website) and ask it to pick a number between 1 and the total value of the whole bill.

That number will indicate a position somewhere down the receipt: if the first four items on the bill cost £10, £9, £11 and £8, and the number is 35, this lands in the middle of the fourth item. For any random number, by adding up the total from the top of the bill downwards, you'll find it lands within exactly one of the menu items ordered.

Then the simple solution is that whoever ordered that item of food or drink has to pay the *whole bill for everyone at the table*. It's easy to administer, and while it might seem unfair on the unlucky person who ends up paying, it's completely reasonable in terms of the probability of who this might be.

The random number has an equal chance of landing on any one of the units of currency spent, and the likelihood of it landing on the menu item of one particular diner is proportional to the amount of money that person spent. It's like buying a number of tickets for a raffle: the more tickets you buy (the more money your meal and drinks cost), the higher chance you have of winning the raffle (getting stuck with a massive bill).

You can take the sting out of this method if you use it repeatedly for the same group of people who often go out for meals together – probability dictates that in the long run they will each pay the bill a roughly proportional amount of the time.

And since the more you spend, the more likely you are to pay, this incentivizes people not to order too much – unless they've got more money than everyone else, in which case they probably won't mind paying. Sounds fair to me!

HOW TO GUESS SOMEONE'S BANK CARD NUMBER

Imagine your friend asks you to get out your wallet and read your bank card number aloud – but tells you to stop before saying the final digit. You play along, and after a moment's pause, they tell you the last digit.

Would you be impressed? Nervous? Maybe both?

This little party trick isn't magic, but rather, the same basic check that's happening invisibly every time you swipe your card, buy something online or enter your card details into a form. Bank card numbers (and many other identification numbers) include a clever self-checking feature that can flag most common errors. The mathematical idea behind it is called the **Luhn algorithm** or the 'mod 10 algorithm'.

CARD NUMBERS ARE NOT RANDOM

Let's try an example, with the 16-digit bank card number:

2364 5981 2406 4139

The first few digits determine what type of card it is, and which bank issued it. The following digits specify your specific account and card. But the final digit doesn't communicate anything like this – it isn't chosen by the bank card issuer, but rather calculated using an algorithm from the previous 15 digits. It's called the check digit, and it allows a check on whether the card number is correct.

In this example, the check digit is a 9. The other 15 digits, sometimes called the 'payload', can be used to calculate the check digit.

THE LUHN ALGORITHM

- Take the first digit of the payload, and double it: if the answer is bigger than 9, then subtract 9 (or, equivalently, add the digits together) to get a single-digit answer, and replace the first digit with that answer.
- Leave the second digit alone. Repeat the step above with the third digit.
- Continue along the number, skipping the even-position digits and doubling/reducing the odd-position digits.
- Now add together all the new digits in your number.
- To find the check digit, find the difference between this sum and the next multiple of 10.

In our example, we start with this payload (ignoring the final 9):

2364 5981 2406 413

Running through the instructions:

- Double the 2. It becomes a 4.
- Move right one digit, and leave the 3 alone.
- Move right one digit. Double 6 to become 12 – this is bigger than 9, so subtract 9 to get a 3.
- Move right one digit, and leave it as 4.
- Move right again and double: the 5 becomes a 10, but this is bigger than 9, so subtract 9, to get a 1.
- And so on...

We end up with:

4334 1971 4406 816

When added up our example gives a total of 61.

The check digit in this case is therefore 9, to bring the total to 70. The check digit will always be the digit needed to get to the next multiple of 10 (which might be zero). This is why the final digit of this fictional card number has to be 9 – if it were any other digit, it would be flagged up as an 'incorrect card number' by card readers or websites.

Every transaction made using a card includes this exact calculation – doubling every other digit, taking the sum, and adding on the check digit. If it gets a total which is not a multiple of 10, then it knows that something is wrong. The complexity of doing different things to alternate digits is necessary so that if a transposition error occurs (such as two adjacent numbers being switched) it will immediately affect the calculation and warn you.

This algorithm is useful as a simple first check of whether you've typed in a number correctly on a website or heard it correctly read out over the phone. All single-digit errors will be picked up, as will most transpositions. You might like to consider which errors will still get through this system.

Luhn's original idea was patented as a mechanical device which required no electricity to operate.

An alternative name for the Luhn algorithm is the 'mod 10 algorithm', which is short for **modulo 10** – which means to ignore everything about a number other than its remainder when divided by 10 (or equivalently, if you're working in base 10, ignoring everything but the last digit). The last step of the Luhn algorithm check uses this idea, which is more generally called **modular arithmetic**.

CASTING OUT NINES

While the Luhn algorithm's final step uses modulo 10, there is a part of the algorithm that feels a bit like modulo 9 – the bit where you add the two digits of a number together, or alternatively just subtract 9.

In fact, if you add together the digits of any number, and repeat this until you have a single digit, you'll end up with the same answer as if you subtracted as many 9s as you could before reaching zero. This is equivalent to dividing by 9 and taking the remainder, and is called taking the number **modulo 9**.

Historically, working modulo 9, or 'mod 9', was used as an error checking method in accountancy – before computer spreadsheets and electronic calculators – because the process of repeatedly adding the digits can be done very quickly with practice. The process is called 'casting out 9s' or 'finding the digital root'.

A full calculation could be done by hand, and the answer could be checked by doing the same calculation with every number replaced by the 'mod 9' version. If these did not give the same result, then you'd know an error had occurred. For example, this calculation adds up 1,459, 2,562, 5,545 and 9,729 with a deliberate, but hopefully forgivable, error.

$$
\begin{array}{rcl}
1\ 4\ 5\ 9 & \longrightarrow & 1 \\
2\ 5\ 6\ 2 & \longrightarrow & 6 \\
5\ 5\ 4\ 5 & \longrightarrow & 1 \\
+9\ 7\ 2\ 9 & \longrightarrow & 0 \\
\hline
1\ 9\ 1\ 9\ 5 & & 8
\end{array}
$$

$$1\ 9\ 1\ 9\ 5 \longrightarrow 7$$

If we turn all the numbers into the 'mod 9' versions, and add them up, we get 8.

Unfortunately, if we cast out nines on the answer 19195, we get 7.

Since the 8 and the 7 do not match, we know that something is wrong. You might like to try spotting where the calculation went wrong. Unfortunately, we don't get information as to *where* the error has occurred by this method (and, in roughly 1/9 of cases, you get the same check digit by a fluke, so it won't pick up the error).

ERROR DETECTION IN THE DIGITAL WORLD

The principle behind the Luhn algorithm, and the process of casting out 9s underpins most modern digital communication – whether it's information being transmitted by a barcode scanner, a text message or a packet of internet traffic. The data is all essentially numerical, and error detection like this can flag up if there's potentially been some information lost or misread – at which point the device can try again.

The next obvious upgrade is an error *correction* algorithm, which can not only notice if there's been an error – it can hopefully fix it too. This kind of technology can be found in many forms of data transmission and storage, including QR codes, CDs and DVDs.

SO...

Even if you can't calculate a missing bank card number yourself, you should at least be thankful that a card payment machine can.

WHAT AMOUNT OF MONEY CAN'T YOU PAY?

In the UK, before 1998, coin denominations available were 1p, 2p, 5p, 10p, 20p, 50p and £1. What's the largest amount you could have had in loose change without being able to pay exactly £1?

This might sound like a strange question, since you'd think the answer would be 99p – above which you'd have £1, and be able to pay it. But if you want to pay an amount exactly – and not receive any change in return – you could have more than £1 and not be able to pay exactly £1. For example, if the only coins you're holding are three 20p pieces, you can't pay exactly 50p no matter what you do.

This kind of question, where quantities exist in fixed denominations, is surprisingly useful, not just when thinking about coins, but for other mathematical problems too.

Imagine you live in a country with only one coin denomination: 5p. You'd only ever be able to pay in multiples of 5p: 10p, 15p, 20p and so on. You'd never be able to pay exactly for anything that cost an amount that didn't end in 0 or 5. (It would also mean everyone had to carry around a *lot* of coins at all times).

But imagine if this country added a second coin, worth 7p. Which amounts would you be able to pay now?

Coins	Total	Coins	Total	Coins	Total
5p	5p	5p+7p+7p	19p	5p+5p+5p+7p	27p
7p	7p	5p+5p+5p+5p	20p	7p+ 7p+7p+7p	28p
5p+5p	10p	7p+7p+7p	21p	5p+5p+5p+7p+7p	29p
5p+7p	12p	5p+5p+5p+7p	22p	5p+5p+5p+5p+5p+5p	30p
7p+7p	14p	5p+5p+7p+7p	24p	5p+5p+7p+7p+7p	31p
5p+5p+5p	15p	5p+5p+5p+5p+5p	25p	5p+5p+5p+5p+5p+7p	32p
5p+5p+7p	17p	5p+7p+7p+7p	26p	5p+7p+7p+7p+7p	33p

Once you get to 24p, there is a run of five consecutive amounts that can all be made using just 5p and 7p coins (the shaded rows). And this means that, if you can make 24p, you just need to add 5p and then you have a way to make 29p – and if you can make 25p you can make 30p, and so on for every quantity above that. The largest number that you can't pay is 23p, and beyond that every quantity is possible.

The mathematics at work here is given the surprising name of **chicken nugget numbers**. In fast food restaurants, chicken nuggets are only available in boxes of specific sizes. For example, boxes might contain 6, 9 or 20 nuggets.

If you wanted 21 nuggets, without throwing any away, you could get two boxes of 6, and a box of 9.

Boxes	Total nuggets	Boxes	Total nuggets
6	6	6+6+6+6+6	30
9	9	20+6+6	32
6+6	12	9+6+6+6	33
9+6	15	20+9+6	35
20	20	6+6+6+6+6+6	36
9+6+6	21	9+6+6+6+6	39
6+6+6+6	24	20+20	40
20+6	26	20+9+6+6	41
9+6+6+6	27	6+6+6+6+6+6+6	42
20+9	29	20+9+9+6	44

For a given number you want to create, and a set of possible units you can make it from, we can use an algorithm to find a combination that works (it's called a **greedy algorithm**, but that's a coincidence and nothing to do with eating a huge pile of chicken nuggets when you're done).

Here's how it works: you sort the units in decreasing order (in the case of our nuggets, we'd order them 20, 9, 6) and then follow the same steps over and over.

- Start by taking as many copies of the largest unit as you can fit into the total, and work your way down the list fitting in as many of the largest ones as possible.

For example, to make the number 48, you could take two 20s, and add one 6 to get 46 – but that isn't the whole amount, and you can't fit any 6s, 9s or 20s in to make up the difference.

So, instead of starting with two 20s, we roll back to a single 20 and try to fill the remaining gap (28) with the next largest unit – 9s. You could fit three of these in, but still have a gap of 1 which you can't fill.

- Repeat this – always trying to fit in the largest unit as many times as possible, and removing one of them if that doesn't work. The full process would be:

$20 + 20 + 6 = 46$ (doesn't work; remove a 20)
$20 + 9 + 9 + 9 = 47$ (doesn't work; remove a 9)
$20 + 9 + 9 + 6 = 44$ (doesn't work; remove another 9)
$20 + 9 + 6 + 6 + 6 = 47$ (doesn't work; remove another 20)
$9 + 9 + 9 + 9 + 6 + 6 = 48$

You could make 48 just using 6s, but the greedy algorithm will try to make it using the fewest items (in this case, boxes of chicken nuggets) possible, so it uses four 9s in place of six of the 6s. It's 'greedy' in the sense that it will use the biggest size boxes it can.

For nuggets in boxes of 6, 9 and 20, the largest number you can't make is 43. 'Largest impossible numbers' like these are called **Frobenius numbers**, and in any situation where a number can be made up by adding together units of fixed sizes, the same maths applies.

In certain sports, points are scored for particular actions in the game: for example, in Rugby Sevens you score 5 points for a try and 7 points for a converted try. This means scores of 3, 6, 8, 9, 11, 13, 16, 18 and 23 are impossible to make in normal play (unless one side is awarded a penalty goal or drop goal, but this is a rare occurrence) – and in the 2014–15 Sevens World Series, no games were recorded where either team had these scores.

Here's a neat way to work out the solution to the coins question from earlier. Start by considering the largest coins first: if we had any £1 coins, we'd be able to pay £1, and if we had two 50p coins, that would also allow us to pay exactly £1. This means we can have at most one 50p coin. Similarly, having five 20p coins would let us pay £1, but four won't let us (and combining any number of the four 20p coins with the 50p still won't give us £1).

Now, if we were to add a single 10p coin, we'd be able to use the 50p and two of the 20ps, so we can't do that – but we can do the same trick we did with 50ps and 20ps again, now with the 5ps and 2ps, meaning we can add 5p and up to four 2p coins and still not be able to make 10p. This means we can have a total of 50p + (4 × 20p) + 5p + (4 × 2p) = £1.43 – and still not be able to pay £1 with any subset of the coins!

SO...

...what amount of money can't you pay? In the UK before 1998 you might have had up to £1.43 and still not have been able to pay £1 exactly. Of course, the introduction of the £2 coin (in 1998) messes up this particular puzzle by making the answer become...infinity. The Royal Mint really should have thought this through more carefully.

CAN YOU GUARANTEE TO WIN THE LOTTERY?

Lotteries are great examples of probability in the real world: everything about them needs to be completely fair, and they are very carefully planned to make sure every number comes up with an equal probability. Lotteries often use carefully monitored machines, lots of cameras and monitoring on the draw itself, and even special sets of lottery balls that are kept safe from tampering.

This means, unlike most real-world situations, the maths can be calculated with almost complete certainty: we know that if a lottery has 50 possible numbers to pick, each one will be chosen with a pretty much exactly 1 in 50 chance.

We can also calculate the probability of particular combinations of balls being drawn – since each time a ball is drawn, the remaining balls are also equally likely. For example, a particular combination of two balls from 50 has a 1/50 chance of the first ball being right, then a 1/49 chance of the second ball being right. For both to happen at once, it's a $1/(50 \times 49) = 1/2,450$ chance.

Of course, for most lotteries, you pick a set of numbers and hope they will match the balls being drawn. In the UK lottery, you pick six numbers from a set of 59, and try to match as many as you can. In this case, the chance of the first one being a match to one of your six numbers is 6/59; then 5/58 for the second, and so on. We can keep going like this, calculating $6/59 \times 5/58 \times 4/57 \times \ldots$ until we have the probability of matching all six balls – and winning the jackpot – which is 1 in 45,057,474.

This is a staggeringly unlikely thing to happen – far less likely than being struck by a meteor (if you stood outside for a year), or dying from being crushed by a falling vending machine (which happens to around two people per year). It's such a tiny probability that it's hard to comprehend, so it sometimes helps to construct situations with the same odds, to see just how small the chance is.

WINNING THE LOTTERY IS AS UNLIKELY AS...

Over all the modern summer Olympic games that have taken place since 1896, 5,793 people have won Olympic gold medals (at time of writing), and probably about 2/3 of these (so, about 4,000 people) are still alive today. There are now roughly 8 billion humans on the planet, so if you were to pick a random person walking down the street, anywhere in the world, there's about a 1 in 2 million chance they're an Olympic gold medallist. This is over 20 times more likely than winning the lottery on a particular draw.

If you instead hoped to run into one of the only 682 humans who have ever been to space, this is still around a 1 in 11.7 million chance: three times more likely than winning the lottery on a particular draw.

A standard mobile phone number has nine digits that can take any value from 0 to 9. A 1 in 22 million chance is about the same probability as dialling a completely random mobile phone number when sitting in a room with 45 people, and it making one of their phones ring. And this is just for one specific lottery: others around the world have much lower winning chances – in some cases, as low as 1 in 200 million.

CAN YOU INCREASE THE ODDS?

Ever since there have been lotteries, people have tried to find ways to gain an advantage, or to somehow cheat probability and do better. But, as we've seen, probability does what it was always going to do, and things happen a very predictable amount of the time, especially when everything is tightly controlled so that it can't be influenced by cheating.

Aside from the ways of actually cheating – creating forged tickets, or attempting to tamper with the draw machine or balls – people sometimes try to use mathematical analysis of the draws to pick combinations more likely to win. This goes beyond having a set of 'lucky numbers', to making actual predictions about which balls are likely to come up in future draws.

For example, people have sometimes become convinced that because numbers appear with a fixed probability, this means they will definitely be drawn by the machine on a regular schedule, and that after a certain number of draws without a particular number showing up, they're more likely – or even guaranteed – to come up on the next one.

While this sounds tempting, it's not how probability works. Every draw is independent, and a ball bouncing around in a machine has no memory of which balls came out before – they still all have an equal chance of coming up each time. On any given 6-ball draw from 59, a particular ball (say, number 17) would have about a 10% chance of being chosen (in fact, it's 6/59 – of the 59 balls, 6 are chosen, and 17 has a 6/59 chance of being one of those chosen).

So while it's very unlikely that the number 17 ball would not be drawn out in a hundred consecutive draws – with a 90% chance of not being drawn in each draw, it's 0.9^{100}, or around a 0.002% chance, in a hundred draws – it's definitely still possible. And, for the following draw, the probability of it coming out remains about 10%.

One way to guarantee you win the lottery jackpot is by buying every ticket. Simply purchasing a ticket for every possible combination of six balls will mean that, whichever six numbers come out, you are guaranteed to win the jackpot.

So what would that involve? Well, given there are 45,057,474 different possible combinations of six numbers to choose from, and that tickets cost £2 each, you'd need to spend £90,114,948 just on buying the tickets. And that's not including the cost of the logistics: just to type the numbers into a lottery machine or website would take a long time – or you'd need to hire a large number of people to help you do it. Even if you can get the time down to ten seconds per ticket, it would take more than 14 years!

The largest ever jackpot in the UK lottery was £66,070,646 – and that was after a long sequence of 'rollovers' with no winner. Since then, this kind of jackpot has become impossible. The current rules dictate that if the jackpot goes over £22 million, it must be won the following week, so if

nobody matches six numbers it's distributed between the people who got five instead.

And this hints at another way in which the plan fails: even if you buy every ticket and guarantee yourself a win, if someone else happens to win too, you'll split the jackpot between you, halving your payout (or worse, if more than one other person wins too). You definitely won't make back as much as you spent!

But it turns out there is a way to guarantee you win something, without spending millions of pounds on tickets. For the UK lottery, prizes are given out for smaller numbers of matches, so matching two or three numbers will win you a small prize. At time of writing, three numbers gets you £30, and two gets you a free ticket to play again, using a randomly chosen set of numbers. If you're prepared to expand your definition of 'win the lottery' to include this, then read on: we have some maths that can help.

A research paper published in 2023 by a group of mathematicians in Manchester, England, explained how it's possible to buy just 27 tickets on the UK lottery, and guarantee that one of them will definitely match two numbers, whatever six numbers are drawn.

The 27 tickets you need to buy are:

1 2 3 4 5 6	9 10 11 12 13 14	18 19 20 21 26 27	32 33 36 37 44 45	46 47 50 51 58 59
1 2 3 4 7 8	9 10 11 15 16 17	18 19 24 25 28 29	34 35 36 37 42 43	48 49 50 51 56 57
1 2 5 6 7 8	12 13 14 15 16 17	20 21 24 25 30 31	36 37 38 39 40 41	50 51 52 53 54 55
		26 27 28 29 30 31	32 33 34 35 40 41	46 47 48 49 54 55
		18 19 22 23 30 31	32 33 38 39 42 43	46 47 52 53 56 57
		20 21 22 23 28 29	34 35 38 39 44 45	48 49 52 53 58 59
		22 23 24 25 26 27	40 41 42 43 44 45	54 55 56 57 58 59

These sets of numbers were devised using a mathematical construction called a **projective plane**. This is a shape which allows you to consider all possible combinations of points and lines for a given set. For example, this shape is a projective plane containing seven points and seven lines (one of the lines is in the form of a circle).

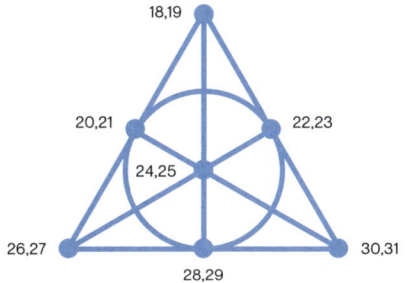

It's called the **Fano plane**, named after Italian mathematician Gino Fano. The important property of this arrangement is that, given any two points, there is *exactly one line they both lie on*, and *each pair of lines meets at exactly one point*. The points are labelled with pairs of numbers from the lottery draw, and you need to buy a ticket representing each of the lines in the diagram – this diagram corresponds to tickets in the third column of the list – and each ticket in that column has six numbers from one of the lines.

The whole system involves five diagrams – three Fano planes like this, and two other diagrams which are simpler cut-down versions of the plane, but with a similar property that each line represents one lottery ticket. Together the five diagrams include a mention of every possible ball in the UK draw (the numbers 1 to 59).

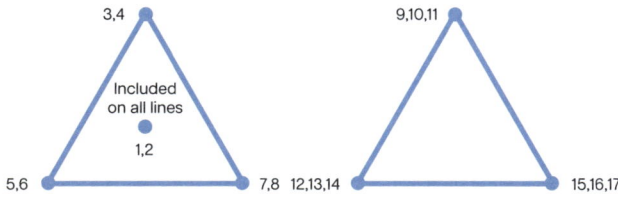

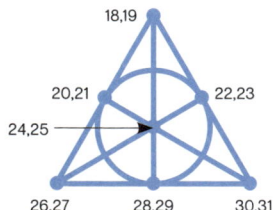
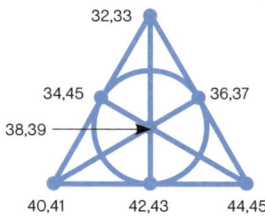
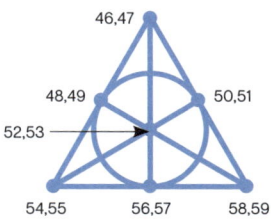

For each diagram, each point is joined to every other point by one line, so if you pick any two numbers that appear on the diagram, they will either be joined by a line or be on the same point to start with. We simply need to buy a lottery ticket for each line, with the six numbers on it corresponding to the ones on that line. To save you counting, the total number of lines in all the diagrams is 27, corresponding to the 27 tickets we listed earlier.

Finally, since there are only five diagrams but they cover all available numbers, the winning lottery draw of six numbers will definitely have at least two numbers from the same diagram (using a piece of maths called the pigeonhole principle, which you'll encounter again later in the book). That means that it will be represented by one of the lines, and, in turn, by one of the tickets you bought. Hooray!

Buying 27 lottery tickets at £2 a ticket will cost you £56, so if all you win is a single free play, you're not going to break even on this (unless that second ticket gets you the jackpot – it could still be you!). But there's no reason why this set of tickets couldn't give you more than one match, or even a match of three numbers. Around a quarter of the time, you'll do better than just a single pair match. Unfortunately, even taking this into account, your expected winnings are still only on average around £20, which represents a loss of £36. Maybe stick to investing your money instead (see page 84 Can you retire early).

SO...

...can you guarantee to win the lottery? Technically, yes. But can you guarantee a profitable lottery win? Sorry...no.

A BET WORTH TAKING?

Are you the betting type? Throughout history, humans have transformed games into gambling, and disagreements into wagers. Whatever you feel about gambling, there's something fascinating about how we attach stakes to confidence – literally and figuratively.

Expressions like 'I bet I can beat you at...!' or 'I bet you're wrong about...!' may well be casual trash talk, but taken literally, they reveal something deeper about someone being willing to risk something on their opinion. The bet amounts (the stakes) are a proxy for how sure they are of their claim. Let's put this idea to the test with a real-world example – an actual bet overheard in a bar:

I bet there are two people in New York with the same number of hairs on their head.

Would you take this bet? If so, how much would you bet?

The more you're willing to risk, the more confident you must be of your answer. So roughly where would you put the chances on the probability scale?

Despite the seemingly vague nature of the question, we can answer this definitively.

AN OBVIOUS WIN

Firstly, let's deal with a small percentage of you who have correctly (and smugly) concluded it is 100% certain, because bald people exist. Well done. You're perfectly correct that there are, and doing what many a true mathematician would do: considering the extreme cases first.

There are definitely at least two bald people in New York, and so if you say they both have zero hairs on their head then you've answered the question.

What's nice about this question though is that, even if we dodge that case, we can still make progress, so let's rephrase the bet as a question:

What are the chances that two non-bald people in New York have *exactly* the same number of hairs on their head?

TWO KEY NUMBERS

We need to make two estimates:

1 How many people are there in New York?
2 How many hairs are there on a person's head?

Both of these seem hard to be precise about, so let's acknowledge that a rough estimate is better than nothing, and proceed on that basis.

1. How many people are there in New York?

This depends what you mean by 'New York', and by 'people', and in fact what you mean by 'in' – but let's make some assumptions. According to 2025 estimates, New York is the most populous city in the USA, with over 8 million people living there. We could take a very conservative estimate of **at least 5 million people**, which would then also describe almost any large metropolis around the world. You will see why we're not overly worried about precision in a moment.

2. How many hairs are there on a head?

This is going to vary between people, but you may enjoy the exercise of trying to estimate it. Some wise advice for this sort of large-number estimation is to consider a smaller version first, and then deal with orders of magnitude.

Roughly how many hairs on a 1cm by 1cm patch of scalp? 1? No, usually more than that. 10? Probably more than that. 100? Maybe. 1,000? Starting to sound crowded.

Let's go with 100 for a 1cm^2 patch as a reasonable estimate.

Now, how big is a hairy head? You might imagine bending a 30cm ruler over your scalp, and it's conceivably the right sort of length to go from ear to ear, and also from forehead to the back of your neck. That would mean your hairy scalp is about 30cm by 30cm, or 900cm^2 in total.

900 × 100 = 90,000 hairs.

That gives us a round estimate of about 100,000 hairs on a head.

Of course, in this day and age you can find answers to questions like this on the internet, and biological estimates vary between roughly 90,000 and 150,000. Apparently, your natural hair colour is a big factor in how densely packed your hair is. Let's go with a generous upper bound and assume the number of hairs on a person's head is **at most 200,000**.

THE PIGEONHOLE PRINCIPLE

The final answer to this problem will require use of an idea that mathematicians call the **pigeonhole principle**. Imagine there are nine pigeonholes, and five pigeons arrive. It would be quite possible for the pigeons to arrange themselves so that none of them have to share a pigeonhole (there are still at least four holes left spare).

If there were nine pigeonholes and ten pigeons, it would definitely be **impossible** that they have their own space. Even if the first nine pigeons all choose a pigeonhole of their own, the tenth is going to have to share with some other pigeon. It's now guaranteed that at least one pigeonhole is being shared.

This is the pigeonhole principle: If there are more things you need to place than there are options for where to place them, you will definitely have at least one place that ends up with two (or more) things placed in it.

It's surprising how powerful this idea can be.

So how does this apply to hair?

- There are at least **5,000,000** people in New York.
- Each of these people has a whole number of hairs on their head, and this will be less than 200,000.
- Since there are at most **200,000** possible different options for the question 'how many hairs do you have on your head?' and 5 million people will each have one of those options, it is **absolutely guaranteed** that at least two of them will share the same answer.

The probability of two people in New York having the same number of hairs is 100% – it definitely must be the case that two people have an equal number of hairs on their head (and probably there are lots more than two of them).

Back to the bet…let's check the phrasing of our opening wager. If a friend said 'I bet there are two people in New York with the same number of hairs on their head,' it would be unreasonable to make a bet that there *aren't*, in fact, two people in New York with the same number of hairs on their head. If anyone disagrees that the pigeonhole principle argument is enough to settle this, the process of actually finding two people and checking if it's true may be less straightforward…

THE NATURAL WORLD

CAN A MONKEY WRITE SHAKESPEARE?

If a monkey was put in a room with a typewriter and left there for an infinitely long time, would it eventually write the entire works of William Shakespeare?

This kind of question is called a **thought experiment**, and it's often helpful to consider situations like this when thinking about knotty mathematical concepts like infinity.

In this thought experiment, the fact that a monkey is involved is largely irrelevant: the main idea is that the monkey serves as a random letter generator, hitting keys on the keyboard with no understanding. This means we can assume that each letter is equally likely to be hit each time. We're also assuming a relatively constant frequency of keypresses, and making the stronger assumption that the monkey doesn't stop typing, or wander off, or decide to throw the typewriter across the room.

The entire works of Shakespeare – 39 plays and 154 sonnets, written over a period of roughly 24 years – contain around 5,327,794 characters, arranged into 884,647 words. Could it ever appear, purely at random, from a monkey's mindless key-mashing?

INFINITY: WHERE EVERYTHING HAPPENS, EVENTUALLY

In fact, given an infinitely long time to type, the monkey is *guaranteed* to type the entire works of Shakespeare.

Part of why this seems so counterintuitive is that we're used to things being finite. But given enough time, and a sufficiently random monkey, every possible string of letters will eventually be generated.

Each time a key is pressed, the probability of a given key being hit by the monkey (including spaces, and a few common punctuation marks) is, let's say, 1 in 30. So, if you're hoping for it to type the word 'bananas', you'll need the first letter to be a B (a 1 in 30 chance), then an A (another 1 in 30 chance) and so on – overall, the probability of typing 'bananas' is $(\frac{1}{30})^7$, which works out to be around 1/21,870,000,000 (1 in 21 billion).

As we saw in the case of lottery draws on page 102, when something has a fixed likelihood of happening (like drawing a particular lottery ball), if we repeatedly try, the probabilities accumulate, and the chances it won't have happened yet get smaller and smaller. But that doesn't guarantee that a thing will happen – each time, there's still a chance it won't (and for typing 'bananas', that chance is very large).

But we have an advantage: our theoretical monkey has an infinite amount of time. We're not just considering the probability of it typing the word 'bananas' starting from the first key it hits: we get another go starting from the second key – if the first key hit isn't 'B', we can effectively start counting again from the next one. So this tiny probability gets slightly larger the longer the monkey types for.

As the length of time the monkey has spent typing gets larger and larger, the probability of it not typing the word 'bananas' gets closer and closer to zero. If we considered a finite length of time, there'd still be a possibility it hadn't happened yet – but since we continue forever, this means that it will definitely eventually happen.

And while the word 'bananas' typed by a monkey would itself be an impressive literary achievement, the same probability logic applies to any finite length of string of characters. Even with the impressive corpus of 5,327,794 characters written by Shakespeare (plus spaces!), infinity is still long enough to guarantee the monkey will eventually type it. (For more on how infinity can lead to slightly strange outcomes, see pages 141 and 148.)

Sometimes this problem is misstated as being about a collection of infinitely many monkeys on infinitely many typewriters; in that case, assuming true randomness in their letter choices (and no monkeys that have particular preferences about which keys they press) at least one of the monkeys would write the entire works of Shakespeare immediately, word for word. In fact, infinitely many of them would. And infinitely many more of them would write an adapted works of Shakespeare where every character's name is replaced with that of a monkey.

THE REAL-WORLD VERSION: NOT SO POETIC

The problem here is infinity. Absolutely any version of this with any *finite* time, and finitely many monkeys, is going to conform much more to your intuition, but as soon as you drop in the concept of infinity, all bets are off.

Infinity is not just big: it's an entirely different concept. If you're attempting a finite thing (like typing a long – but definitely finite – string of 5.3 million specific characters) infinitely many times, it's going to happen in every possible way, and each of the finitely many ways it could happen will each occur infinitely many times.

A 2024 research paper in this area studied the 'Finite Monkeys problem', considering what would happen if a finite number of monkeys were given a finite amount of time to write the works of Shakespeare. And this is where the importance of infinity becomes clear – researchers Stephen Woodcock and Jay Falletta considered this finite case and concluded that, since the universe we live in has a finite lifespan, even if every monkey on Earth was given a typewriter they wouldn't by chance write out the works of Shakespeare before the universe itself ended.

Indeed, assuming a world population of around 200,000 chimpanzees working simultaneously for the roughly 10^{100} years before the universe is predicted to end, the researchers calculated that, while there's a pretty good chance one of them would hit on the word 'bananas' (in fact, it's likely enough to be almost certain), the probability of one of them typing all of Will's 884,647 words before the heat death of the universe was around $6.4 \times 10^{-7,448,254}$. That's a probability of one in a number that has seven and half million zeros on the end.

This is why the original thought experiment used an infinite time period – without that detail, the question isn't particularly interesting. (It also makes some assumptions about the longevity of monkeys – if we're restricting the problem to the real world, we should probably remember that monkeys need food, and that they'd need to take breaks, and we'd need to replace them when they succumbed to old age.) This is why it's a thought experiment, and hopefully nobody has tried to conduct it for real!

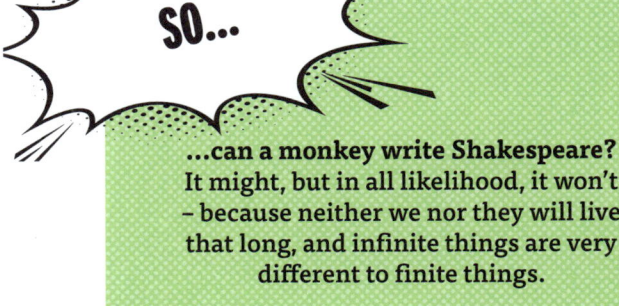

SO...

...can a monkey write Shakespeare? It might, but in all likelihood, it won't – because neither we nor they will live that long, and infinite things are very different to finite things.

WHY DON'T RABBITS RULE THE WORLD?

Biologists care about how many individuals are in a population of animals, and how this changes over time. Some populations die out, some stabilize and some tend to grow, or behave wildly unpredictably. In the 1970s, scientist Professor Robert May showed how a mathematical model could capture all of these outcomes in one relatively simple formula. What determines whether a community of rabbits explodes in size to frightening numbers, whether they're happily stable or whether they all die out?

In the natural world animals will breed and they will die, either through old age or lack of food, or from predator attacks. May, among others, proposed a mathematical model based on two simple starting ideas:

- **REPRODUCTION:** Animals reproduce by breeding, and the more animals there are, the more breeding will occur.
- **STARVATION:** Animals die, partly because the available food will feed only a limited number of animals, and if there are too many animals there may not be enough food to go around. The closer to the limit, the more starvation occurs.

We can think of the population growth as a step-by-step process, with a value n that increases as time passes. Given the current size of the population at year n, what will it be after one year (year $n+1$)?

Let's model a rabbit population. The current size of the population is denoted x_n, and the size the following year is x_{n+1}. These values are both numbers between 0 and 1, where 1 is the largest the population can possibly be, and 0 is no rabbits at all. The new population, at the next point in time, can be calculated using a formula called the **logistic map**:

$$\underset{\text{Next year's population size (relative to current population } x_n)}{x_{n+1}} = r \times \underset{\text{Reproduction}}{x_n} \times \underset{\text{Starvation}}{(1-x_n)}$$

Here, the amount of reproduction that happens is related to the current size of the population: the more rabbits there are, the more reproduction will occur. So we include the current population size (x_n).

For the starvation part of the equation, there'll be some maximum population that the current amount of resources will support. The closer we get to that, the more rabbits will die of starvation. The formula has a term for $(1-x_n)$, which will be very small when the population is close to full – meaning we're multiplying by a number close to zero, reducing the size of the population.

CHANGING THE REPRODUCTION PARAMETER

At the front of this formula is a number, or parameter, r – this can be thought of as a **reproduction rate**, partly determined by how fertile the animals are (famously, rabbits reproduce more quickly than, say, pandas). In general, the higher this number, the more the population has a tendency to reproduce.

For small values of r (up to 1), you'll find the population dies out: for whatever starting value of x_n you choose, after a small number of steps the population drops to zero, representing the rabbits dying out.

For a value of r between 1 and 3, the population tends to stabilize: they reproduce at the same rate as they die out, and the number of rabbits stays constant. The higher the value of r, the larger this population will be – for example, at $r = 2$, the population sits around half the maximum possible size.

Going beyond $r = 3$, things start to get a bit more interesting. Here, the population is reproducing quickly, but this leads to a large population – too large for the available resources – in turn leading to a much lower population, due to starvation. From $r = 3$ to around $r = 3.4$, the population flips between two values each year (often called a 2-cycle).

TRY IT AND SEE

You can see the effects of changing r for yourself using a scientific calculator (you'll need one with an 'answer' button – often labelled *ANS*).

- Enter a starting population of 0.5 by typing 0.5 and pressing equals.

- Pick an r-value between 0 and 4.

- Type in the formula, using *ANS* instead of x_n, so that the calculator uses the previous answer in the next iteration of the formula. i.e. [*your r-value*] × *ANS* × (1−*ANS*)

- Press equals a few times and observe how the population changes over the years.

Be prepared for very different outcomes depending on your choice of r.

Real-life animal populations with high fertility often exhibit this kind of behaviour, causing a repeated cycle of frantic breeding followed by overpopulation and starvation.

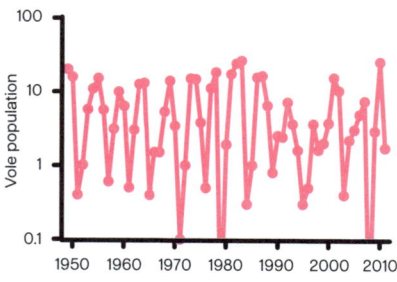

Real data of vole population estimates

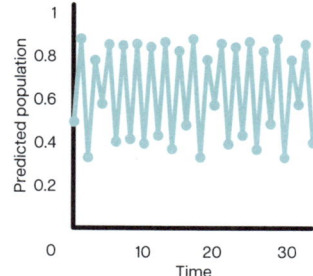

Output from a logistic map model with $r = 3.59$

At $r = 3.5$ the outcomes start to change quickly: 4-cycles, then 8-cycles, then 16 and so on, until outcomes with no clear pattern start to appear at around $r = 3.57$. This type of behaviour is often called **chaotic** (and this model, called the **logistic map**, was one of the very first systems seen to exhibit 'chaos' in the modern mathematical sense).

A typical sign of chaotic behaviour is that a tiny change in the starting conditions (the initial size of the population) can lead to very different behaviour – within the chaotic region, we see everything from stable 3-cycles (as at $r \approx 3.83$) to essentially random and apparently unpredictable changes in population size.

This relatively simple formula can be used to model many observed real-world biological outcomes, and it also opened up a mathematical 'can of worms' – kickstarting a new discipline in mathematics called **non-linear dynamics** (and the study of mathematical chaos – for more on this, see page 122).

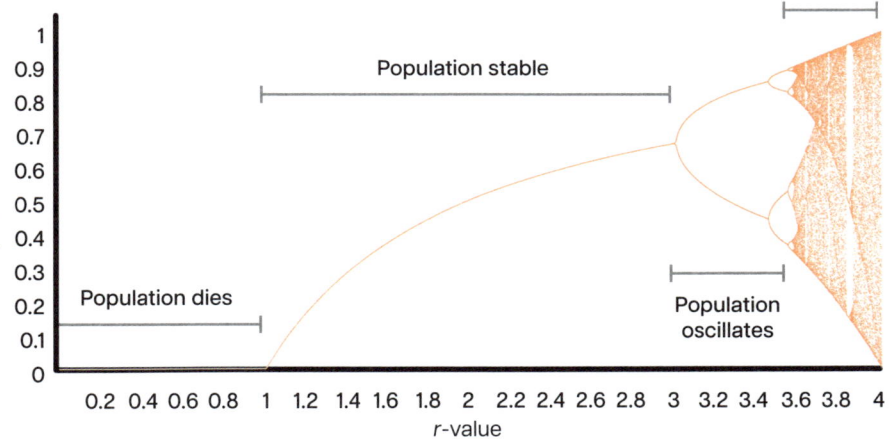

The long-term outcomes for populations with different r-values

For populations that we can model in this way, the number of animals usually stays within a reasonably small range – due to food availability, as well as the inherent instability that comes into the system when reproduction rates get too high. This means that – even though they're literally at it like rabbits – maths tells us that rabbits will never quite manage to take over the world.

CAN A BUTTERFLY CAUSE A TORNADO?

You may have heard of the 'butterfly effect' (and not just in the context of a few questionable films with that title). You may also have heard of the 'domino effect'. While they're both used idiomatically to talk about cause and effect, the difference in original meaning between these two phrases is a mathematical story worth hearing. Let's talk about chaos.

Many people use both these phrases interchangeably, to mean that a small initial input can start a chain reaction to cause a huge eventual output. The 'domino' part of the phrase comes from the idea of standing up lots of dominoes before knocking the first one down and causing a long chain of toppling dominoes.

Normally, all the dominoes in a standard set are the same size. But it's possible to topple dominoes of increasing size, so that a tiny domino starting the chain causes an arbitrarily big domino to fall by the end of the chain. This is a classic piece of physics fun – but what's it got to do with butterflies?

The 1995 and 2004 films with the title *The Butterfly Effect* also deal with the unforeseen chain reactions of consequences of someone's actions, and there is a reference to the idea in Ray Bradbury's 1952 time-travel short story 'A Sound of Thunder', where the death of one butterfly causes far-reaching consequences for the future.

Modern cultural references aside, the 'butterfly' association almost certainly comes from a now-famous 1972 scientific conference session, presented by Professor Edward Lorenz and based on his research into mathematical weather prediction. It had the title 'Does the flap of a Butterfly's Wings in Brazil Set Off a Tornado in Texas?' He knew what he was doing in using such a dramatic-sounding title – even back then 'clickbait' had its uses – but he was making a serious point, which had almost the opposite meaning of the 'domino effect'.

His main thrust was not that a tiny butterfly flap might cause a dramatic effect like a tornado – but rather that a tiny change in the starting flap might result in a vastly *different outcome*. For example, the butterfly flapping a wing in one direction might eventually lead to a storm; whereas a flap the other way might lead to a complete absence of any storms.

CHAOS IS NOT RANDOMNESS

What Lorenz was trying to say was that a system like the weather might be *extremely sensitive to the initial conditions* – so much so that an almost insignificant change in the starting conditions may result in wildly different eventual outcomes (even though they're all still weather forecasts). Part of Lorenz's point was also that this sort of surprising sensitivity might be more widespread than anyone previously expected.

This might sound familiar from the idea of population dynamics outlined on page 120 – Lorenz also used the logistic map to make this point, before Robert May popularized its application in population modelling.

Lorenz, working with computer scientists Ellen Fetter and Margaret Hamilton, discovered this idea during research in

1963 on modelling the weather with numerical simulations. They had developed a set of differential equations which successfully modelled a simple weather system as it evolved over time.

$$\frac{dx}{dt} = \sigma(y - x)$$
$$\frac{dy}{dt} = x(\rho - z) - y$$
$$\frac{dz}{dt} = xy - \beta z$$

The set of differential equations that Lorenz, Fetter and Hamilton used (now called the Lorenz Equations)

Without going into detail, suffice it to say that the equations were solved numerically by giving a computer starting values for the variables and iterating to get the next values predicted over time.

The team discovered an interesting problem: they had a printout of some previous calculations, and Lorenz decided to repeat the calculations to see what happened if he let them go further. Rather than starting from the beginning, he typed in numbers from the previous printout.

Crucially, these numbers had been rounded from 6 decimal places to 3 decimal places on the printout to save space. Lorenz went off for a coffee while the computer repeated the previous calculations (or so he assumed). When he returned, he found the computer had produced a completely different weather prediction from the previous run.

On closer examination, it appeared the output had started very similarly to the previous run, before diverging to a very different outcome.

The team assumed that there were problems with the code or the hardware. So it took some time to eventually realize that the only actual difference was the seemingly insignificant changes to the starting points, which came from the typed-in numbers with a few digits rounded off.

This was a revelation. This extreme sensitivity to changes in the starting values is part of what mathematicians now call chaos, or chaotic behaviour. We saw it in the logistic map modelling populations, and Lorenz's team saw it in the equations they were using to model the weather.

It's now an established property of mathematical models of all sorts of phenomena like fluid flow and biological systems. This is a big reason why predicting these things in the long term remains difficult. In general use, the word 'chaotic' means something that's completely unpredictable and doesn't follow any rules – but mathematical chaos follows the rules of the system exactly, yet appears to generate randomly different outcomes, depending on what your exact original input is.

WHY WEATHER FORECASTING IS HARD

We now know that predicting the weather, or the behaviour of any chaotic system, is only possible in the short term – not because long-term prediction is too hard for our mathematics, but because the mathematics itself contains an inherent tendency to become unpredictable. Forecasting becomes a battle to extend the short window of predictability before chaos takes over.

The precise definition of chaos remains a point of contention even among mathematicians, but Lorenz himself summarized the essence of it as follows: 'Chaos: When the present determines the future but the approximate present does not approximately determine the future'.

There is another pleasing butterfly coincidence to leave you with here. The Lorenz system of equations contains three variables (x, y and z). When you plot any solution of the equation with these three variables over time on a 3D graph you see a remarkable picture.

While it's true that any tiny change in the starting conditions causes a wildly different outcome at a given time in the future, the pattern of all the outcomes over time seems to settle into a similar shape – now called the 'Lorenz attractor'.

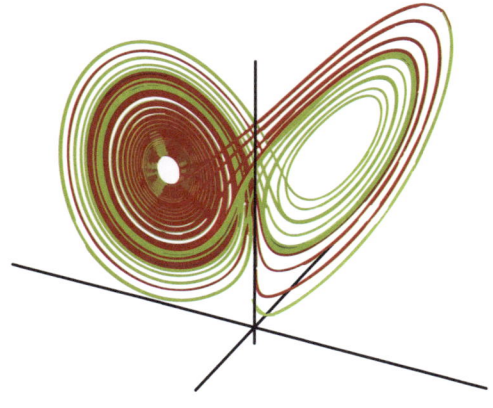

The Lorenz attractor – the solutions to the Lorenz system always take this shape, even though they are still extremely sensitive to the initial conditions

It looks extraordinarily like a butterfly – two 'wings' of points, with the results whirling around one wing or the other over time, and occasionally jumping between them.

This is a pure coincidence, and many other different 'strange attractors' of various shapes exist – often intricate and fractal-like in structure (see page 141) – but the fact that this first example looks like a butterfly is too good to ignore.

SO...

...can a butterfly cause a tornado? No, not really. But can tiny changes be important? Yes, absolutely, and so much so that in any chaotic system no achievable increase in precision can outweigh their eventual effect. The butterfly effect is real, important and sometimes beautiful.

WHY IS THERE A HEART IN MY TEACUP?

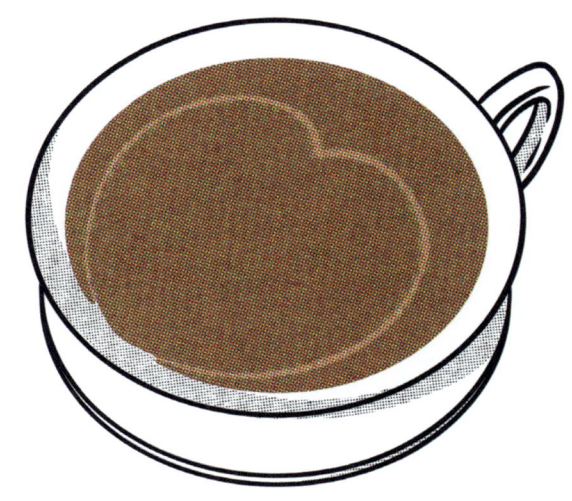

Have you ever noticed an intriguing heart-shaped pattern in the light dancing on top of your cup of tea, or perhaps at the bottom of your cup? It's often more noticeable where there's strong morning sunlight, or a bright spotlight in a kitchen ceiling.

If we're forcing some romance on this, it's a heart shape. If we're being more basic, it's a bum shape. If we're being more mathematical, it's a **nephroid**, a **cardioid**, a **caustic**, or an **epicycloid**, or some combination of these terms, depending on specific conditions.

A LIGHT-HEARTED EXPERIMENT

Before we get into the detail you might like to try recreating this yourself. A setup like the one in this diagram will give you a good chance to see the effect – particularly if your light source is the only one around, and your mug is white or reflective. You can experiment with moving the light source around to see different effects.

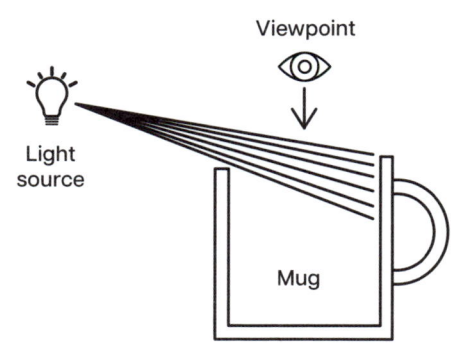

What you're seeing is the result of the light, which famously always moves in straight lines, reflecting off a curved surface. In this case, the curve is part of a circle (your mug wall). While this doesn't stop the light from moving in straight lines, it does mean that the reflected rays then overlap in curious ways.

The bright shape you see is where many of the reflections overlap, making it look brighter than where the rays don't overlap. This shape is called a **caustic** – which originally comes from the Greek, and then Latin, word *causticus*, meaning burning. (You might recall the effect that sunlight can have when focused through a lens.)

It's the same effect that's happening when you see shimmering bright patterns at the bottom of shallow water in the sun. But why does it create a heart shape in the teacup?

CURVE STITCHING

Imagine ten light rays coming in from a long way away – this is pretty much true if they're from the sun – so they're effectively parallel. When they hit the far wall of the mug they will reflect like this:

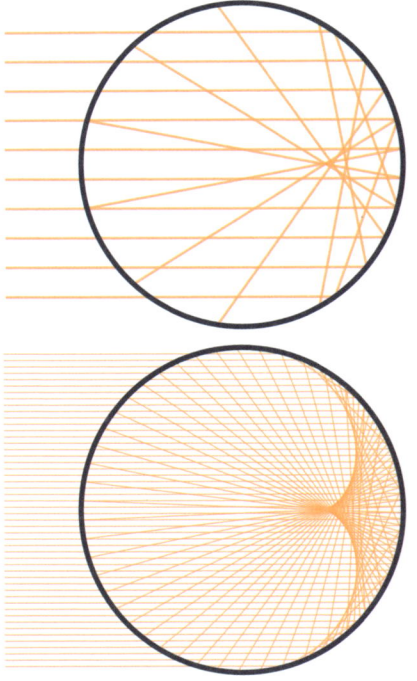

The pattern may not be not obvious yet, but if we increase the number of lines simulating the light to 50, we start to see a familiar shape:

Unless you use a light source made from a single beam like a laser, light tends to shine in continuous blocks, so cranking up the number of lines starts to better simulate the reality of light. This diagram has 1,000 lines, and looks more realistic.

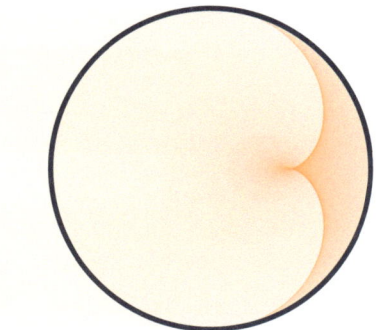

The curved shape that becomes visible is formed by the reflected lines overlapping on that particular curve. You may remember the effect of creating curves from straight lines from school exercise book doodles. If you put some equally spaced dots round a circle, number them, then connect each dot to the one 3 times bigger (overlapping round the circle as needed) you will create a curved shape.

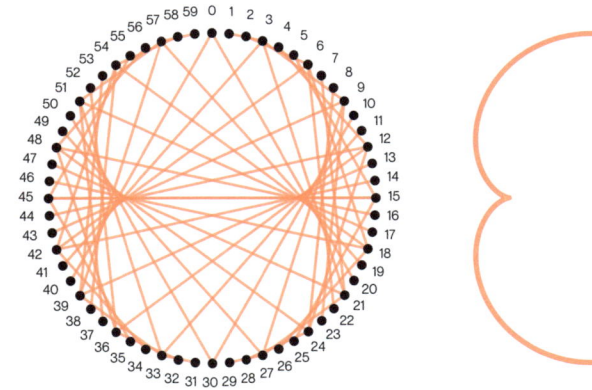

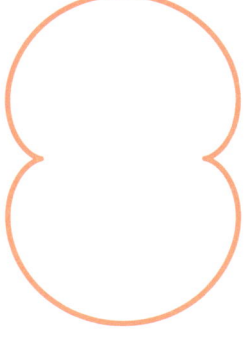

This shape has become known as a **nephroid** (Aristotle joined the two Greek words for 'kidney' and 'shape' to describe what he saw). A nephroid is one particular example of a type of curve called an **epicycloid** – in this case, one with two cusps (the pointy bits).

'Wait, this is not a heart shape!' I hear you cry. If you feel cheated by the title of this section, worry not – the shape changes if you move the light source (as you may have noticed if you played with your own version).

Putting the light closer to the mug – or actually on the edge of it – makes the shape shift to become what's known as a **cardioid** (literally 'heart-shaped'). In this diagram, there's a nephroid overlaid in blue, which traces the same curve as the cardioid down the right-hand side. The cardioid is another kind of epicycloid, but with only one cusp.

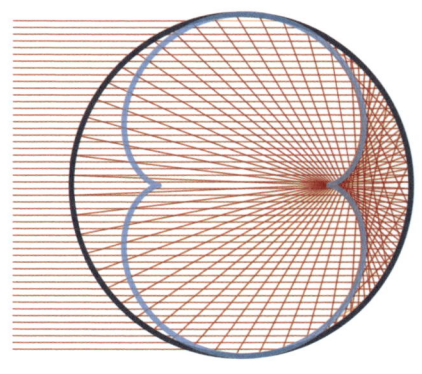

Mathematically, these shapes can both be described by parametric equations:

$$x = k \cos(t) - \cos(kt)$$

$$y = k \sin(t) - \sin(kt)$$

for $0 \le t < 2\pi$.

Using $k = 2$ we get the cardioid, and $k = 3$ gives us the nephroid.

For a fun distraction you can try typing this into a piece of graph plotting software (such as GeoGebra or Desmos) and changing k to some new values, to see some of the rest of the family of epicycloid curves.

Or, for a different way to create curves, you can use the same straight-line doodle technique from earlier. Connecting numbers on a circle to those which are three times bigger gave us the nephroid, and connecting numbers to those which are twice the size will give the cardioid.

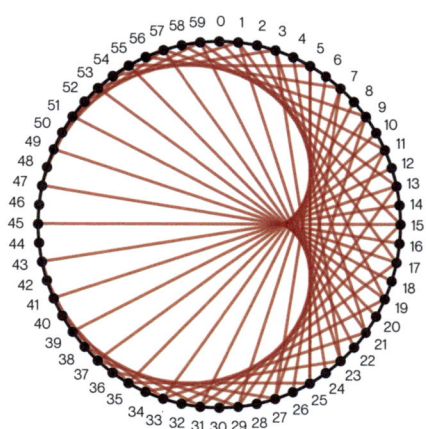

Before we get too lost in the detail, remember that we're always imperfectly modelling the real situation. For example, we haven't taken into account that the light has to come down at an angle, so the curve it hits is actually not a perfect a circle, and rather an ellipse. This and other nice upgrades on the model have been discussed by mathematician Dr Dominika Vasilkova.

ON FURTHER REFLECTION...

So far so good, but you may be reflecting (haha) on the fact that once the light has got into your mug (or other circular shape) it seems likely to keep bouncing around. You would be right, of course – but each time another reflection occurs, the light gets fainter, which is why the shape you see is likely to resemble the nephroid or cardioid shapes we've discussed; these are the shapes produced after one bounce.

But it's natural to ask what happens with later bounces. If we draw all the lines that formed the nephroid, and reflect them one more time, we get a new caustic shape. In this diagram the original nephroid is in orange, and the second set of reflections is in pink. This time the shape formed is much more like a heart, but we're in territory without well-defined names now.

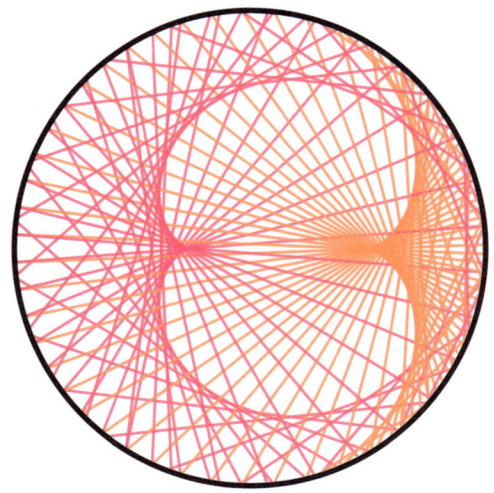

It can be trickier to see this in home experimentation, but you might manage it with a shiny gold ring, lit from the side in a dark room.

Finally, it would be a sad day if nobody asked 'what happens if you keep going?' – a healthy mathematical habit – so here's a picture of the third-order reflection (in green, on top of the previous two reflections). It creates the most heart-shaped image of the lot, so far.

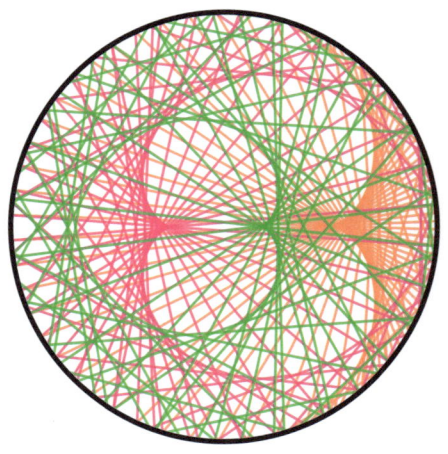

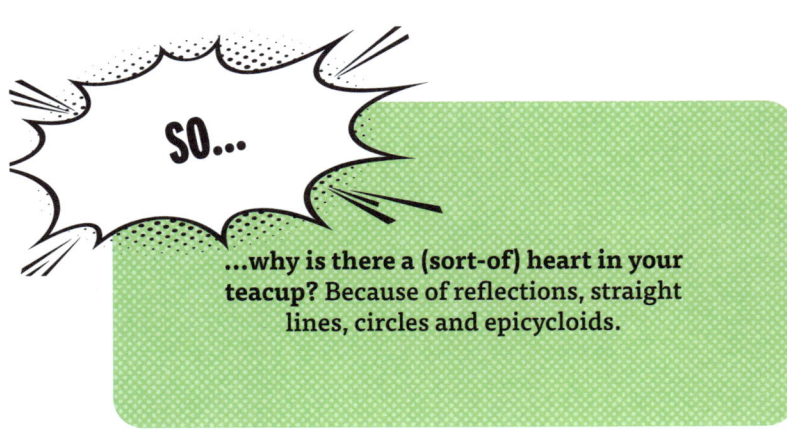

SO...

...why is there a (sort-of) heart in your teacup? Because of reflections, straight lines, circles and epicycloids.

WHY DO SUNFLOWERS SPIRAL?

Imagine you're a flowering plant. You want to grow in such a way as to maximize the benefit you get from your environment – for example, to distribute your growing leaves to efficiently fill the space around your stem, but still access the sunlight you need for photosynthesis. Or to grow your seeds so that they efficiently fill the space around your flower head.

The goal is the same in both these cases – you want to use the space efficiently – and the biggest decision you have to make is where to grow the next leaf or seed. This comes down to a question of angles. Of course, flowers are not actually 'thinking' or 'deciding' this sort of thing: evolutionary pressures select for more efficient behaviour – this is how natural selection works. But if you were going to design a leaf or seed arrangement yourself, how would you do it most efficiently?

Let's consider the best way to arrange seeds in a flower head (but remember this might apply just as well to arranging leaves around a stem).

FLOWER ARRANGEMENTS

Consider the central point of your flower head (the black cross in the circle in the diagrams), and grow a seed out in a particular direction (off to the right). If you continue to grow seeds out in exactly that direction you'll end up making a long straight line of seeds. This is a waste of the rest of the space, and it's obviously inefficient.

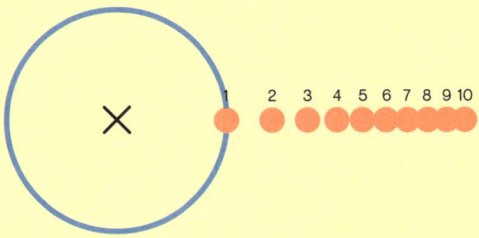

Can we do better?

What about growing a seed, then changing the direction you're generating seeds in, by, say, *half* a turn, before growing the next seed. Repeat. Congratulations, your seed head now has two lines! But it's still inefficient, and a far cry from any flower seed head you've seen.

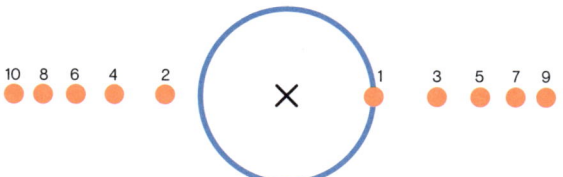

Proportion of full turn between seeds: 0.5

Now let's try one third of a turn between each seed. This gives three spokes of seeds. And if we turned a quarter turn, we'd get four spokes.

It might be becoming obvious that there's a connection between the fraction of a turn that the flower uses, and the number of spokes created: if you turn $1/n$ of a turn each time, you'll get n spokes.

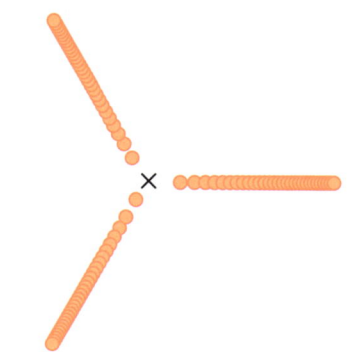

Proportion of full turn between seeds: 1/3

Sometimes different fractions give the same arrangement of spokes.If you turn 1/5 or 3/5 of a turn, you still get five spokes. The denominator (the number on the bottom of the fraction, when the fraction is in its lowest terms) determines the number of spokes.

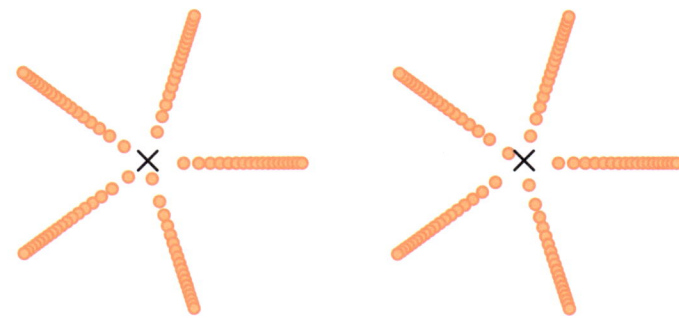

In the image on the left, the proportion of a turn between seeds is 1/5; on the right it is 3/5. They end up looking very similar.

Interesting effects start to appear when you turn through angles that are not described by fractions with small numbers on the top and bottom. For example, 11/23 of a turn produces this:

There are 23 spokes visible in the outer regions, as you might expect with a denominator of 23, but the inner region clearly has a spiral with two intertwined spiralling 'spokes' – and this is because 11/23 (= 0.478...) is quite close to one half, or 1/2.

Proportion of full turn between seeds: 11/23

None of these seed arrangements are a particularly efficient use of space though. In fact, spokes are exactly what we'd like to avoid, because they have lots of empty space between them.

Can we find a fraction of a full turn that doesn't produce spokes? Yes! But to do this, we need to use a 'fraction' of a turn that is not a fraction. Any number that can be written as a fraction is called a **rational number**, because it is a **ratio** between two whole numbers. To dodge the spokes, we need a number that **cannot** be written as a fraction – it must be a non-rational number, usually called an **irrational number**. It won't have a denominator, and we won't get spokes! Or, so you'd think.

IRRATIONAL SEEDS

There are many famous examples of irrational numbers: one example is π (pi), sometimes called the 'circle constant', used in calculations involving circles. Since it's irrational, that means $1/\pi$ is also irrational – so what does the picture look like for $1/\pi$ of a turn?

Proportion of full turn between seeds: $1/\pi$

Weirdly, there are still spokes, but they are slightly curved. There are 22 of them. This is because, although π is definitely an irrational number (it can't be written as a ratio, which is why we use a Greek letter symbol to describe it), it is very well approximated by a ratio: $22/7 = 3.142857...$ So $1/\pi$ is well approximated by $7/22$, and that's what's causing the 22 spokes. (You might notice another approximation to π in the centre of the diagram: $\pi \approx 3$, giving three spokes at the centre.) It also gives visual evidence for an otherwise somewhat baffling claim: π *is not very irrational*.

Some other irrational numbers do better. For example $1/\sqrt{2}$:

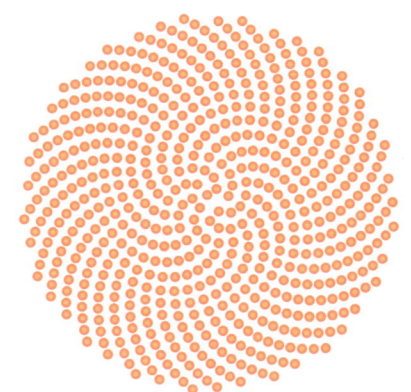

Proportion of full turn between seeds: $1/\sqrt{2}$

This time we stop seeing such obvious spokes and the seeds feel more efficiently spread out. That said, you might be able to convince yourself that there are 41 (very curved) spokes visible at the outside edge. And you'd be correct, because $29/41$ is a close approximation to $1/\sqrt{2}$.

So $\sqrt{2}$ is 'more irrational' than π, in the sense that it's less well approximated by a ratio (and so looks less 'spokey' in our image).

Since some numbers are more irrational than others (in some sense), we can look for one that is more irrational than all the others. In this context it must relate to a seed arrangement that is the most 'un-spoke-like' possible. Sounds like an efficient candidate for flowers to use...To cut a long story short, it's this number:

$$\frac{\sqrt{5}-1}{2} \approx 0.6180339887...$$

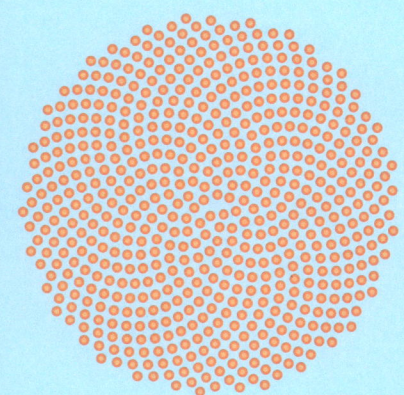

Proportion of full turn between seeds: $\frac{\sqrt{5}-1}{2}$

THE MOST IRRATIONAL NUMBER

We're back to a number we've met before in this book, although you may not have spotted it yet. This is the golden ratio (see page 35). Whatever else is sometimes claimed about the golden ratio, it is an important number, and here is a glimpse of both its mathematical claim to fame, and the source of the claims that it is somehow a number of 'nature'.

The angle associated with this arrangement of seeds (about 0.618 of a full turn) is approximately 137.5°, and often called the **golden angle** (this is the angle that successive seeds will appear at in our flower head).

As discussed earlier, the golden ratio is defined as $\frac{\sqrt{5}+1}{2}$ $\approx 1.6180339887...$ and often given the symbol φ (the Greek letter phi). The number we used here is the reciprocal $1/\varphi$, to make it a proportion of a turn – but, by its very definition, φ is the number that has the same decimal expansion after the decimal point when you take its reciprocal.

There are technical mathematical justifications, beyond just looking nice in these images, why this is (in a sense) the most irrational number. But, even without going into detail, there's another level of commonly claimed overlap here.

We found the golden ratio earlier by taking successive ratios between pairs of Fibonacci numbers, and looking at the limit of that process. If you check the image again, even though there aren't well-defined 'spokes', you may be able to count spirals. The image we've used here has

55 spirals running in the clockwise direction, but you may also be able to make out spirals that run anticlockwise – there are 89 around the outer edge. These numbers are not a coincidence: they are numbers in the Fibonacci sequence: 1,1,2,3,5,8,13,21,34,**55**,**89**,144,…

Using the golden angle is the most efficient way to pack things that grow at successive angles. And the golden ratio is increasingly well approximated by pairs of numbers in the Fibonacci sequence. This is why the natural world ends up frequently appearing to 'choose' Fibonacci numbers of spirals in spiralling growths.

We should remember that nature is not 'choosing' to use these patterns – it's just being efficient, usually by the process of natural selection forcing inefficient organisms to die out. The study of how plants arrange their seeds and leaves is called **phyllotaxis**, and the mathematics we need to describe the efficiency we see in phyllotaxis is the mathematics of irrational numbers, and in particular the 'most irrational' number – the golden ratio. And that's why sunflowers spiral.

HOW LONG IS A COASTLINE?

Geographers are often interested in measuring the distances between different places. But imagine you've been asked to measure the length of the border of a particular country. How might you approach this?

BORDER MEASUREMENTS

You could grab a world atlas and measure the edge of the shape of the country with a ruler. But the outline of a country is not often a straight line, so you'll be forced to fit your ruler along a lumpy, bumpy shape to get a rough measurement, and work your way around the outside. Once you've counted how many times your ruler will fit around you can convert this to an real-world length using the scale of the map. For example, if the ruler is equivalent to a length of 20km, and we can fit 11 copies of it around the edge of the country, this gives a rough estimate of 220km for the border length.

If you flip to the page for that specific country you may find a larger-scale map, which you can try to measure again – and you might find you get a slightly different answer. The version of the map on this page is more zoomed in, so your ruler now corresponds to a shorter distance in kilometres.

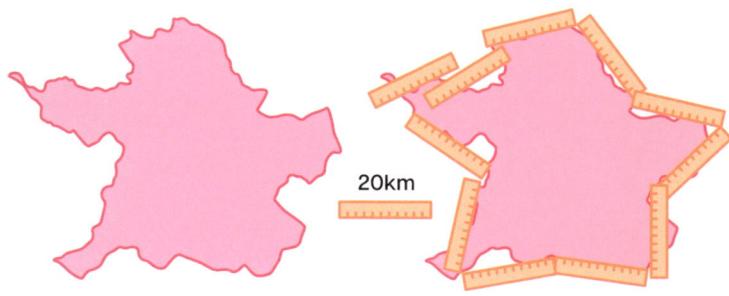

20km

11 sections, each 20km: total = 220km

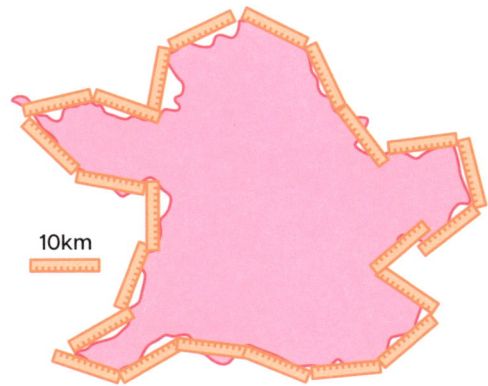

10km

23 sections, each 10km: total = 230km

The edge of the country has more detail at this scale – it wiggles in and out more, and your relatively shorter unit of measurement means you can capture more of that detail – so your measurement of the length will almost certainly be longer.

In fact, the shorter the unit you use, the longer the length of the border you'll find. With 10km units, you can fit 23 of them around the map, giving a total length of 230km.

If you went down to the level of measuring the coastline by trying to count each tiny grain of sand, the length would be incredibly long – it's still the same distance from one end of the coastline to the other, but the line you're measuring the length of wiggles in and out much more as you reduce the length of the unit of measurement. You can capture more detail at smaller scales, so the length you can measure is longer.

This happens because shapes with complexity like this, which occur in nature, have a property that means they're like a **fractal** – a mathematical shape that's defined in a simple way, but ends up being infinitely complicated and detailed.

DEFINING FRACTALS

Imagine we start with a straight line that's 27cm long. We can draw this on a normal piece of paper with a ruler.

Then we apply one simple rule to the line: *replace the middle third with a pair of lines forming two sides of an equilateral triangle.*

Our 27cm straight line would have a 9cm middle third, and we'd replace it with a inverted V-shape with sides of length 9cm. Now our line has four 9cm sections, meaning its length is 36cm overall.

Then we follow the same rule again: take each straight section of line, find its 3cm-long middle third, and swap it for a 3cm V-shape.

This gives us a line with 16 straight 3cm sections, giving a total length of 48cm. We can keep repeating this same step, and each time the line will get longer, and more wiggly.

Eventually we reach a shape called a **Koch curve**, named after the mathematician Professor Helge von Koch. This shape is an example of a mathematical structure called a fractal. In order to be a true fractal, we need to continue repeating the same simple step forever – repeating it infinitely many times.

This should give you a clue that it's not possible to construct a true fractal – in the real world, the limits of physics dictate the smallest something could be, so it wouldn't be possible to draw

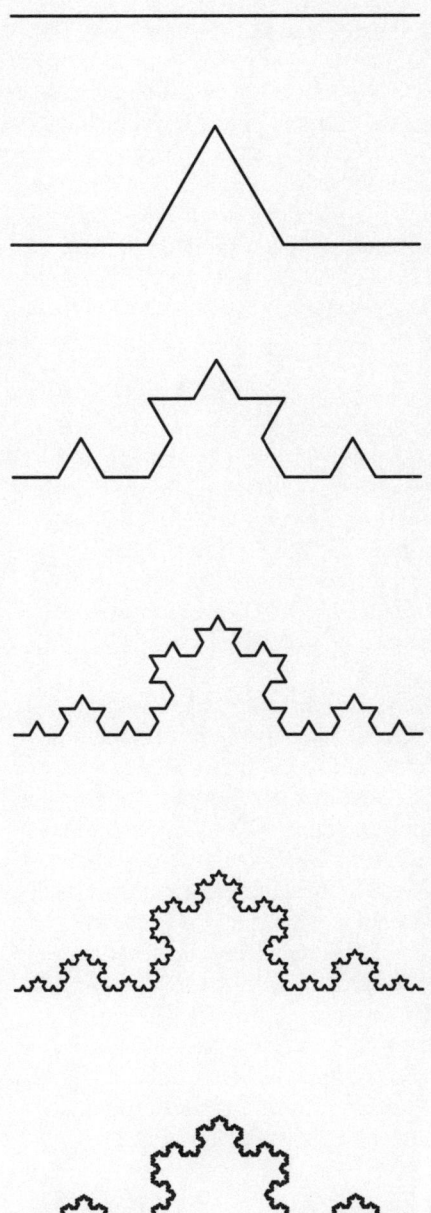

a line shorter than 10^{-35}m (called the Planck length). It would also take an infinitely long time to construct, and unless you're an immortal monkey with a typewriter (see page 114), nobody has that long to spend drawing a wiggly line. That said, it's not particularly different from the claim that you can never construct a true circle – there will always be imperfections – and yet the *idea* of a circle remains clear, and is pretty useful. The same is true of fractals.

You can generate fractals in many ways – by removing or replacing segments from a straight line, by cutting out pieces of 2D or 3D shapes, by sticking together multiple copies of a shape and then scaling it down, or by using mathematical functions that dictate which individual points are inside the fractal and which are outside.

But while fractals come in many different forms, they have certain properties in common. One is that they have **self-similarity** – they contain smaller copies of themselves or have parts that look like the whole shape. They also have **structure at every scale** – no matter how far you zoom in on a fractal, there'll always be some part of it that has some interesting structure, like a corner. (Compare this with a shape like a square, where if you zoom in on the edge, it becomes a straight line and remains a straight line no matter how closely you look.)

Fractal tree, constructed by splitting each branch into two branches

Sierpiński triangle, constructed by removing the central triangular section from each triangle

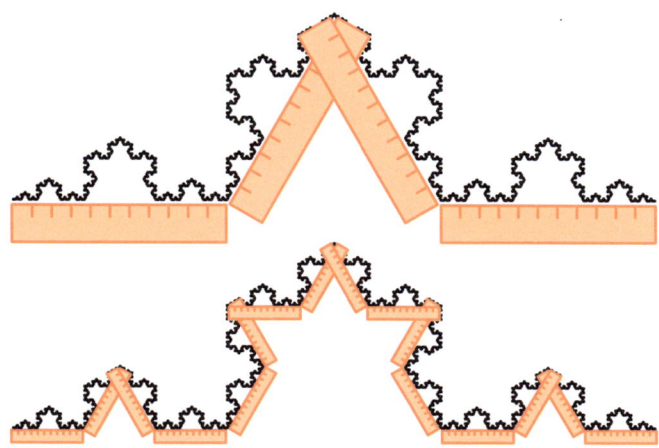

With a 9cm ruler the measurement of the length of the Koch curve is
9 × 4 = 36cm; with a 3cm ruler it is 3 × 16 = 48cm

The Koch curve is a true fractal: it contains many smaller
copies of the same structure repeated within the shape, and
has infinite levels of detail. The length of the line gets longer
with each step – so it could be considered to be infinitely
long, even though it fits into a space of 27cm on a page.

But it could also be considered as having a length which
varies depending on how accurately you measure it: your
measurement of the length depends on how many of the
layers of tiny triangles you measure.

Like with coastlines, the measurement you get for the length
of this fractal line between two points depends on the length
of your ruler. And coastlines have some other properties in
common with fractals too.

Here are two jagged lines, both of which
represent a section of coastline. One of them
is about 20km, and one is around 350km long.
Can you tell which is which?

Coastlines are often considered 'fractal-like' in nature. They don't have the same kind of infinite detail as our Koch curve, but they do have a form of self-similarity – sections of coastline, when zoomed in, look similar to coastlines on a larger scale.

These two sections of coastline with different levels of zoom have the same kind of structure – to the point that it's difficult for us to tell them apart. (If you're curious to know, the top one is a short 20km stretch on the north coast of Spain near Santander, and the bottom one is a large chunk of the south coast of Spain from Barcelona to Valencia, around 350km long.)

Mathematicians have developed precise ways of measuring how strong the fractal property of an object, including a coastline, is, and this is often referred to as the **fractal dimension**. While we normally consider objects to be one-dimensional (like a straight line) or two-dimensional (like a square), fractals – and other objects with fractal-like properties – can have numbers of dimensions that lie in between two whole numbers.

For example, the coastline of Norway (famously wiggly, because of all the fjords) has an estimated fractal dimension of around 1.52, whereas the coastline of the UK has an estimated fractal dimension of about 1.25. The Koch curve from earlier has a fractal dimension of 1.26, which is calculated as log(4)/log(3) since, at each level of scale, it contains four copies of something three times the size.

So while fractals don't really exist (other than in our imaginations), natural shapes do often contain fractal-like structures – like the Romanesco broccoli, which looks like a plant straight out of a maths textbook!

And how long is a coastline? Well, that literally depends on the length of your ruler...

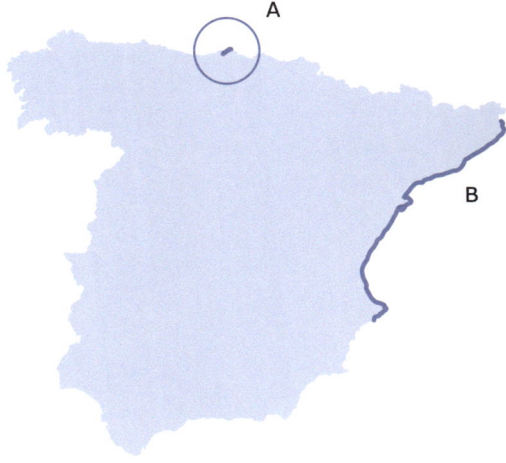

ZENO'S PARADOX

The Greek philosopher Zeno posed a series of paradoxes, many of which concerned the nature of reality, measurement and time. The most famous is the paradox of Achilles and the tortoise, which involves a race between the fast runner Achilles and a slow-running tortoise.

The tortoise is given a 100m head start on Achilles (which seems only fair, since Achilles runs much faster) and both set off at the same time, heading in the same direction. You might imagine it'd be easy for Achilles to leave the tortoise in the dust, given how much faster he can run. But Zeno argues that not only can Achilles not beat the tortoise in the race, he won't even be able to catch up with it!

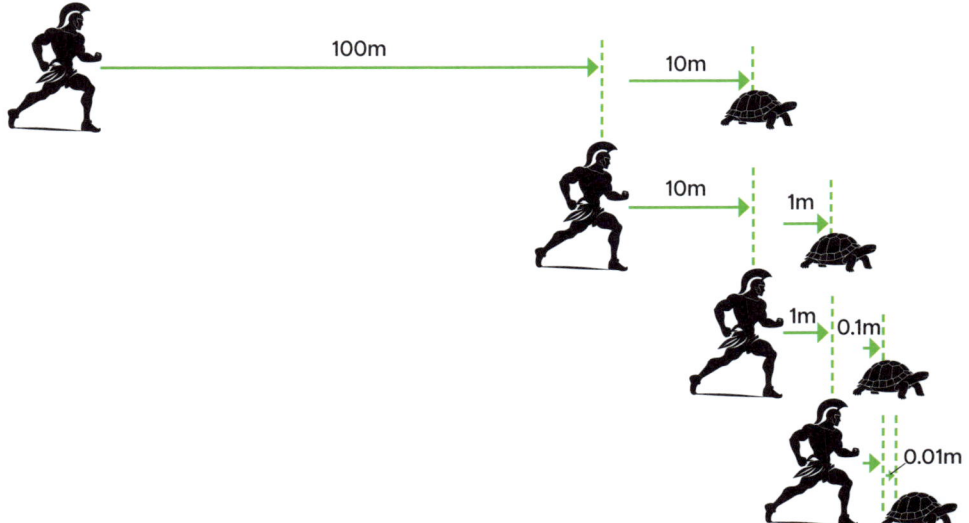

Zeno's argument goes as follows:

(For our example, let's assume Achilles is ten times quicker than the tortoise.)

- After an amount of time has passed, Achilles will reach the 100m mark, where the tortoise started from. At this time, the tortoise will have moved a much shorter distance: 10m from where it started.

- After another amount of time, Achilles will now reach this 110m mark, where the tortoise just was – but by the time he gets there, the tortoise will have moved again.

The argument can be repeated again from here – each time Achilles reaches the point where the tortoise previously was, it will have moved from that spot. So can Achilles ever reach the tortoise, or indeed overtake it?

The problem Zeno faced when making this argument was that the mathematics of his day (c. 450 BCE) didn't yet have the tools to understand what's happening. Logically, Achilles should definitely be able to run past a slowpoke tortoise; and at the same time, Zeno's statements are definitely true. So what's going on?

Imagine if the tortoise was on a rocket skateboard, and was able to match Achilles' speed exactly. In this situation, when Achilles reached the 100m mark, the tortoise would be 100m ahead – and it definitely would be impossible to catch. You'd have two objects moving at the same speed, one of which is 100m ahead of the other.

But in this case, the tortoise is moving at one-tenth of the speed of Achilles. So his first section is the initial 100m head start; then a 10m section to get to where the tortoise was when he was at 100m; then a 1m section, then a 10cm section and so on.

Zeno's argument was that since we're always adding some distance on the end, we'll never reach the point where Achilles overtakes the tortoise. But these distances are getting shorter every time, and in a specific way: it's called a **geometric series**, because each length is the previous one multiplied by a constant factor (in this case, 1/10).

Even though our sequence goes on forever, if we add the lengths, we find ourselves getting closer and closer to a limit:

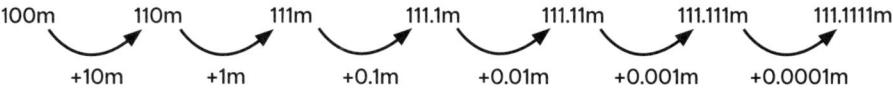

| 100m | 110m | 111m | 111.1m | 111.11m | 111.111m | 111.1111m |
| +10m | +1m | +0.1m | +0.01m | +0.001m | +0.0001m | |

We could carry on adding 1s to the end of this number forever – in maths, we say it's an infinite series, defined by adding an infinite list of numbers together. The numbers in a series are called 'terms', and in this series each term is 0.00...01m bigger than the previous one (with an extra zero before the 1 each time).

Each time we add a term, it puts an extra '1' digit on the end of the decimal expansion – and if we were to carry on doing it forever, we'd have an infinite string of 1s going on forever, giving exactly 111 and one-ninth, denoted 111.1̇m.

We call the number at the end of the infinite adding process the **limit** of the series – in this case, the limit is 111.1̇m. So the tortoise would be overtaken by Achilles after they'd run exactly 111.1̇m, which is a precisely defined point, a little over 111m away from the start line.

Zeno argued that, because reaching this point involved infinitely many steps, it somehow doesn't make sense to ever finish. The mathematical insight he misses is that it's completely possible to add together an infinite list of numbers and get an answer that's a finite number – provided the numbers you're adding together are getting smaller fast enough.

This still proves surprising to many people, depending on how they first encounter it, and formal mathematics didn't really settle on the language and tools to understand it until into the 19th century, when the ideas of calculus and convergence became more rigorous.

Achilles does indeed have to run the length of each of these infinitely many sections of track, but each one takes him one-tenth of the time (and effort) it took him to run the previous one, and beyond a certain point the remainder of the sections would all blur together into one rapid succession leading up to the instant he waves at the tortoise on his way past.

Infinite series like this, which add up to a finite total, turn up all over the place. Anything which grows in a way that means each term is a fixed smaller proportion of the previous one (our example had one-tenth) will have a finite sum.

The series where each term is half the previous, starting from 1, adds up to a total of exactly 2.

$$1 + \frac{1}{2} + \frac{1}{4} + \frac{1}{8} + \frac{1}{16} + ... = 2$$

This means everything past the initial 1 adds up to another 1. You can see this by looking at a diagram like this:

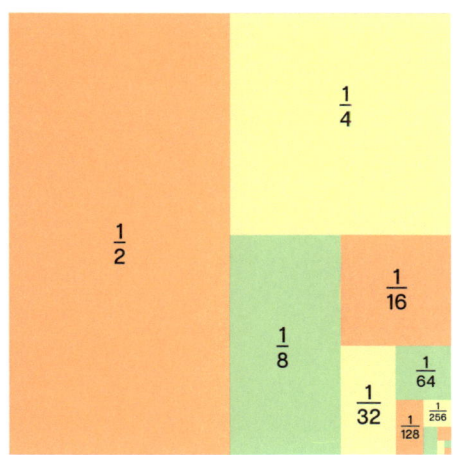

The whole square represents one unit; and, within, it contains a half, a quarter, an eighth and so on. Each of these takes up half the space the previous one did, and each time there's room to use the next term to fill half the remaining space.

Any finite series, adding up some but not all of these numbers, and stopping after a certain number of terms, wouldn't fill the whole of the box – but if we carry on to infinity, the limit of the series is exactly 1.

This is the origin of a classic maths joke:

Infinitely many mathematicians walk into a pub.

The first orders a pint, the second orders a half-pint, the third orders a quarter of a pint, the fourth orders an eighth of a pint...and so it continues

The bartender looks down the infinitely long queue, thinking about how long it's going to take to pour so many drinks. Then, suddenly realizing what's happening, they quickly pour two pints, give them to the mathematicians to sort out, and shout, 'The problem with you mathematicians is: you need to know your limits!'

PLAYING TO WIN

WILL WE SEE A ZERO-MINUTE MILE IN 2560?

Take a look at this sporting data:

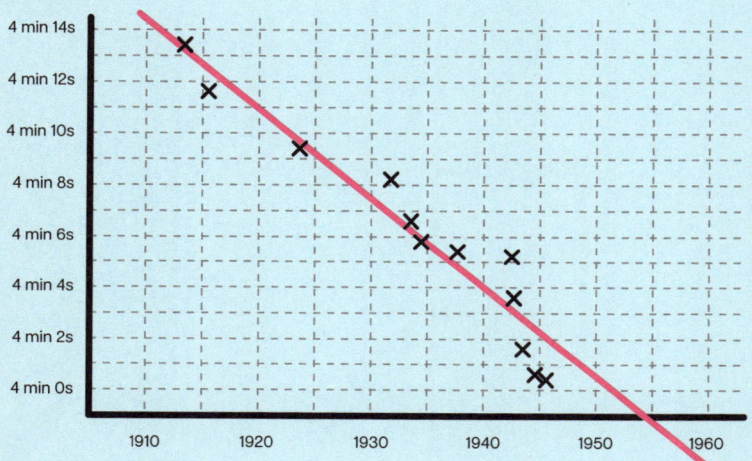

Can you guess what it's measuring?

Now let's set the scene: it's 1950, and the official world record for a man running a distance of one mile is currently at 4 minutes and 1.4 seconds, and has been for nearly five years. Nobody has ever run a mile in competition in under 4 minutes, but the sporting world is beginning to sense that it might just be possible. Well done if you correctly guessed that the data represents the men's world record times for running one mile.

The graph gives pretty compelling evidence for the excitement in the running community. If you draw a 'line of best fit' (mathematicians would call this a **linear regression line**) it would probably look something like the line on the graph.

If the trend continues, it looks like someone will break the 4-minute record before the 1950s are out. The word 'if' is doing some heavy lifting here, but this sort of reasoning is ubiquitous, and sometimes justified. The trend line crosses the 4-minute barrier around October 1954 – and while in 1950 no-one knew what was going to happen, from our more privileged view in the future we can see that history justifies this prediction. On 6 May 1954 Roger Bannister ran the first official sub-4-minute mile, in a time of 3 minutes 59.4 seconds.

So what does this tell us about trends in sporting data? Can we use them to make these kinds of predictions?

PREDICTING THE FUTURE

Here's the graph extended to include the data of men's one-mile world records up to the present day.

The blue line shows the updated trend line, taking into account the full datatset including post-1950 records.

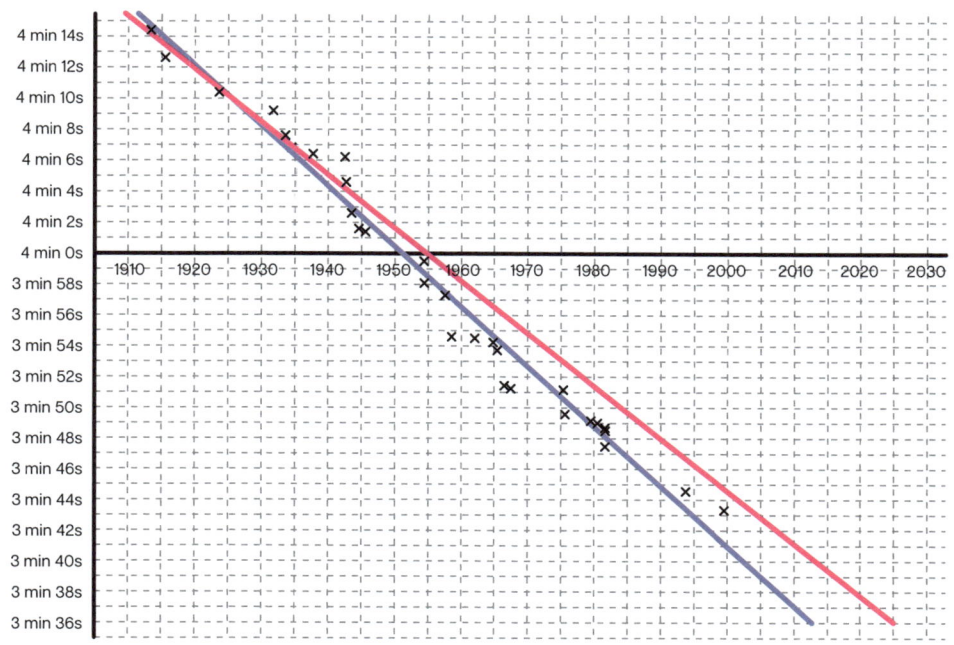

This shows an impressive trend: the men's world record for one mile has become steadily faster at a rate of about 4 seconds per decade (that's the gradient of the trend line) for almost 80 years.

But both of these trends have missed something crucial. No-one (at the time of writing in 2025) has beaten the world record set by Hicham El Guerrouj in July 1999. The gap on the right of the graph, between this 1999 record and the present, is just as informative as the data itself – the gap is telling us something, but the trend lines, because of the way we draw them, will never pick this up.

Extending a trend line beyond any dataset is asking for trouble, unless there's a particularly good reason to justify why that trend line is actually caused by something, and not just a statistical fluke.

In this case, the trend line is pretty good for all the data we have, right up until 1999, then it blithely predicts that the world record will have tumbled down to about 3 min 36s by 2025. Not many people would reasonably place a bet that the next world record for a man running one mile is going to be as low as 3 min 36s.

There are many mathematical lessons to learn here, but one is really really important:

Extrapolation is dangerous.

Another tool we can use instead of extrapolation is 'interpolation', meaning using the trend line to predict values between data points in the dataset, and it's often (rightly) given much more statistical credibility. It's much easier to argue about a missing data point when you have data points either side of it to back you up. In this case, though, interpolation is a bit nonsensical too – since it's claiming that there 'should' have been world records being broken, at about that rate, in gaps where no-one was breaking them. That's not how records work.

It is a sobering lesson to push the extrapolation all the way to its nonsensical end. Here's a plot with the women's mile world record times too, in green, with a linear trend line.

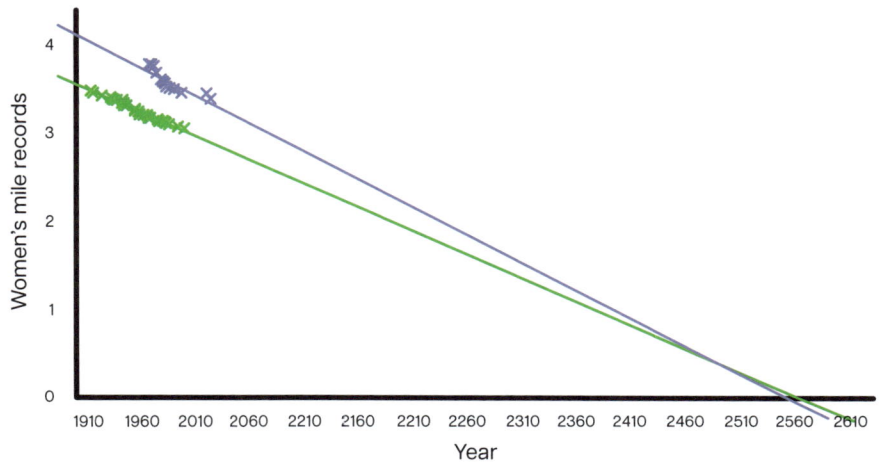

Pick your favourite headline:

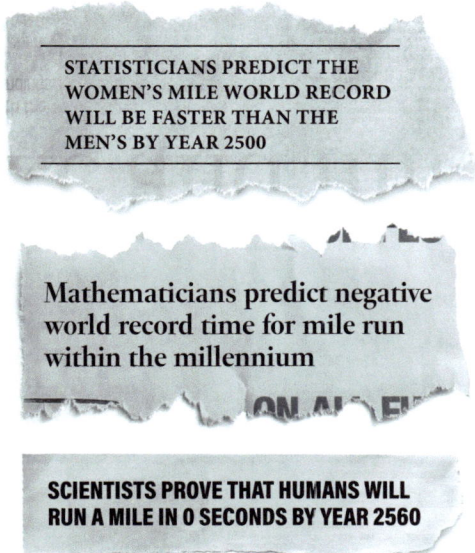

STATISTICIANS PREDICT THE WOMEN'S MILE WORLD RECORD WILL BE FASTER THAN THE MEN'S BY YEAR 2500

Mathematicians predict negative world record time for mile run within the millennium

SCIENTISTS PROVE THAT HUMANS WILL RUN A MILE IN 0 SECONDS BY YEAR 2560

Obviously, no sensible statistician would make these predictions based on this data, but the reality is that even the most rational humans make this sort of statistical error all the time, and it's worth being on your guard.

IF THE CURVE FITS...

Let's finish with a final cautionary tale. It's always possible to fit a curve to data points: it is a solved problem, and the ability to do so is built into any spreadsheet (even if the points are all over the place). For example, take four random points:

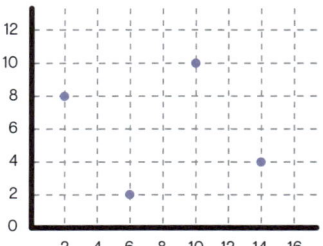

There is an exact curve that goes through all of them. It happens to be a cubic curve (the equation describing it has an x^3 term in).

But just because this *can* be done, doesn't mean it *should*.

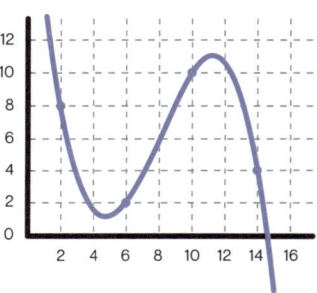

In early May 2020, at the height of the COVID-19 pandemic, the Trump administration in the US were quoted as preferring to use a 'cubic model' of the death toll due to the virus, and that this model predicted that 'deaths essentially stop by May 15'.

Matthew Yglesias (for the Vox.com website) looked into what they might have meant by this. Here's what his scatterplot of data on the daily number of deaths over the first few months of the pandemic looked like.

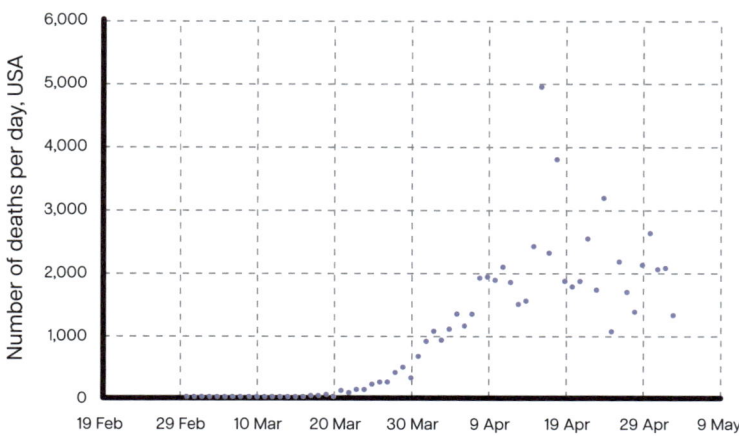

Any spreadsheet can find a line of best fit for this data (though it's arguable that this is already a terrible idea):

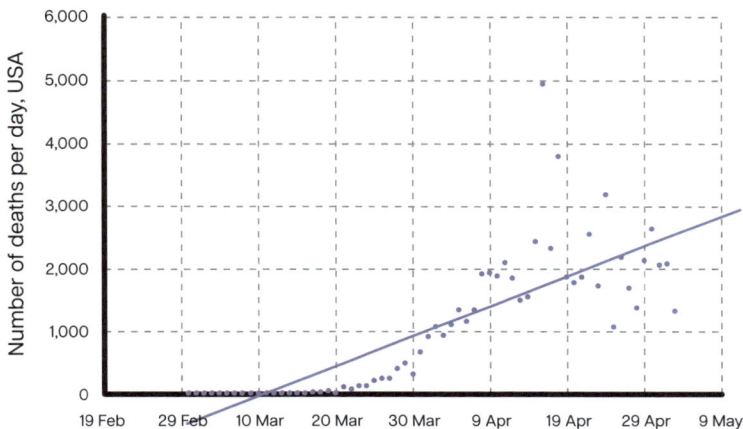

But it's just as possible to fit a different curve model to it – perhaps a 'cubic model'?

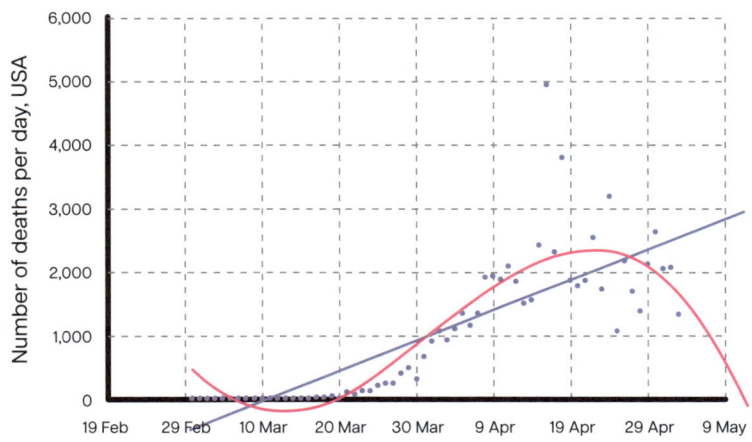

Hey presto – the trend makes it look like the deaths will stop by around 15 May.

But they didn't.

At the end of May 2020 around 1,000 people were still dying each day from COVID in the US, and by March the following year around 500,000 Americans had died in total, according to the COVID Tracking Project (which collected data up to early March 2021).

Unless we have a good reason to suspect a trend line is accurately capturing the behaviour of a dataset, extreme caution should be used if extrapolating from it. With the right choice of model, you can make it say whatever you like – and this is why mathematicians need to think about where that model comes from, and why the data is doing what it's doing.

Modelling the real world is difficult – but mathematics gives us some tools to make predictions, which when applied carefully can be very powerful. But we're not likely to see a zero-minute mile by 2560.

WHERE'S THE BEST PLACE TO TAKE A RUGBY KICK?

In the game of rugby, you immediately score some points when you score a 'try' – like a touchdown in American football, when you manage to touch the ball onto the ground behind the opponent's goal line – but then you also get the chance to take a *conversion kick*.

This is where you try to kick the ball between two posts and above a crossbar (the posts are the big H-shaped things you may have seen on rugby pitches). The name 'try' originally came from the idea that it would allow you to 'try kicking for a goal', which you might then 'convert' into points. Confusing terminology aside, this is the big idea in all variants of the sport of rugby.

After the try is scored, you have a decision to make – you are allowed to kick the ball from anywhere in a direct line, parallel with the sidelines, down the pitch from where you scored the try.

WHERE SHOULD YOU KICK FROM?

If the try is scored near the centre of the pitch (under the posts) then the decision-making process is fairly straightforward. The further back you go, the longer kick you'll have to make, and the narrower the gap between the posts will appear.

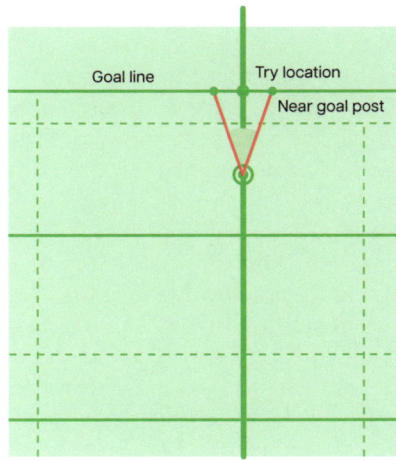

So, in this case, you go as close as you can, to give yourself the widest possible target to aim at. In practice, players usually choose a spot that's a few metres back from the goal line, to give them space to clear the bar.

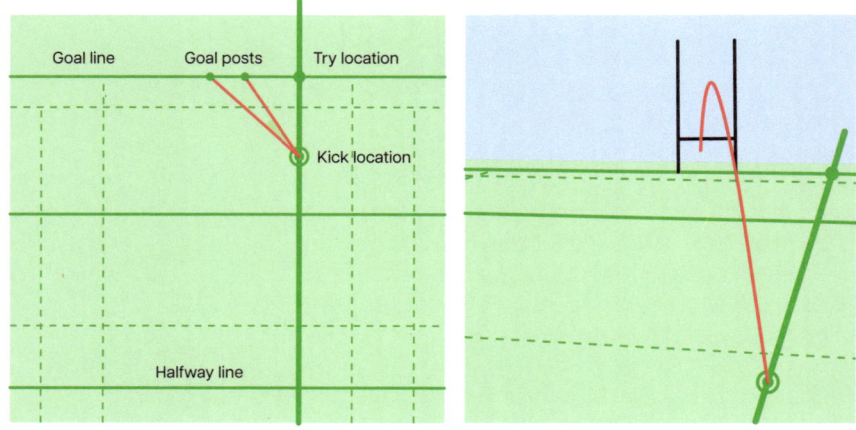

Kicking from a position close to the goal, as in the left-hand image, gives an obviously tight angle. Kicking from a position further away, as in the right-hand image, gives a narrower gap to aim for.

If you score the try to either side of the posts, then things are less obvious. If you go too close to the try line, even though your distance is now short, the goal posts will appear extremely close together – leaving a very narrow gap visible, and requiring a very accurately directed kick. Then again, back off too far, and the posts will be very far away, again giving you a narrower gap to aim for, as well as a longer kick.

There must be a sweet spot somewhere: imagine walking backwards from your try location, while staring at the goal posts. At the start the visible angle is zero (when you're on the try line). As you walk down the pitch the angle from you to the two goal posts opens up, but eventually starts to look narrow again as you get further away.

So what's the best spot to pick? In practice, players tend to use their intuition, taking into account the distance they can kick, and how big the target looks. But using a bit of mathematics, we can calculate what the optimal position is.

Imagine the two goal posts and a potential try-kicking spot are points on a circle. The line from the try location down the pitch (the one that the player has to pick their kicking spot from) will cross the edge of the circle twice, at the two points marked x in the first diagram. One of these is the potential kicking spot, and the other is the next spot along the line

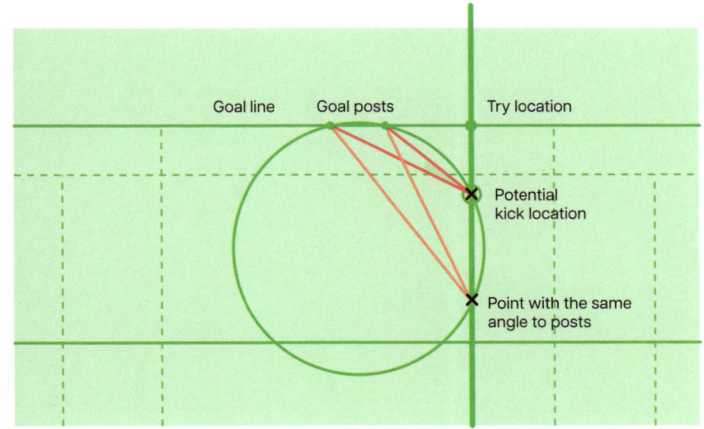

Goal line · Goal posts · Try location

Potential kick location

Point with the same angle to posts

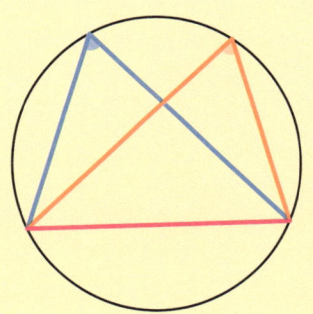

CIRCLE THEOREM: ANGLES IN THE SAME SEGMENT ARE EQUAL

The two angles (blue and orange) formed from the same chord (the pink line) in the same segment (the portion of the circle above the pink line) are equal.

where the gap between the posts looks the same – thanks to the circle theorem quoted in the box (the 'chord' is the goal line between the posts, and the two angles both come from this chord).

If we move backwards from the *closer* kick spot, the visible angle increases (the circle on which the two goal posts and the kick spot lie gets smaller as we do this). The visible angle also increases if we move forwards from the *further* away point. There must be a point between them which gives the largest possible visible angle. At this point, the circle between this spot and the goal posts is *just touching* (mathematicians would say 'tangential to') the line running back from the try location.

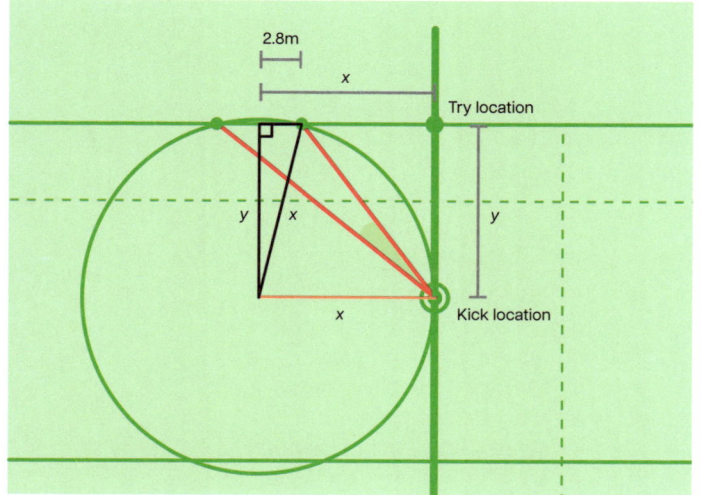

Best spot

The optimal
kick spot for
this try location

So where will that optimal spot be in general? With one more
diagram and a bit of maths we can finally answer this question.

Define x to be the sideways distance from the centre of the
posts to your try location (this is the bit you will already
know, based on where you scored, and will also be the radius
of the circle), and define y to be the optimal distance back
from the goal line to take the kick from (this is the bit you
want to know). We also need to know that the posts are
normally 5.6m apart, so the distance from each post to
the centre of the goal post is 2.8m.

2.8m

x

Try location

y x

y

x

Kick location

The geometry
of the best
kick spot

We can use this information to draw a right-angled triangle with sides 2.8 and y, and hypotenuse x (since it's also a radius of the circle). Pythagoras' theorem gives us the fact that:

$$x^2+y^2=2.8^2$$

We'll know x from our try location, and we want to know y, so we can rearrange this to make y the subject:

$$y=\sqrt{x^2-2.8^2}$$

This is the equation of the curve that represents the best place to take the kick from. It's the orange curve on the diagram below. For any particular try location, you should walk back until you reach the orange curve to optimise your goal kicking angle. It's doubtful that rugby players are using this formula in their head mid-game, but you might notice that the orange curve is very close to the pink line, which is a 45° line from the centre of the goal.

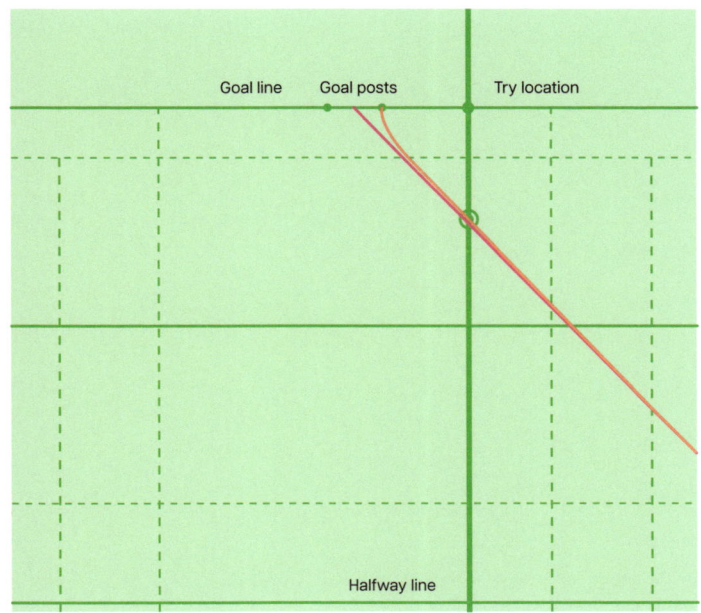

Goal line   Goal posts   Try location

Halfway line

The orange curve is another example of a hyperbola (one of the conic sections – see page 61), but the useful thing about a hyperbola is that it gets closer and closer to a straight line the further you go along it. You can see this in the equation – the 2.8 becomes increasingly irrelevant as x gets large. If the 2.8 were not there at all, this would be the simple formula $y = x$ (a 45° straight line).

In practice, this means you'll be close to the optimal goal angle if you simply pace back exactly as far from the try line as your try spot was from the middle of the pitch (so the angle to the goal is 45°).

There's still a question of range to worry about – if you can't kick very far, then you may well sacrifice some goal angle to get a bit closer. But like all modelling exercises, this is a good starting point from which to refine your model with the details and subtlety that real life presents us with – is it windy, are you tired, and so on.

The best place to take a rugby kick – for you – is wherever you feel confident about hitting the target from; but some basic geometry suggests you should take it from a line about 45° from the centre of the goalposts.

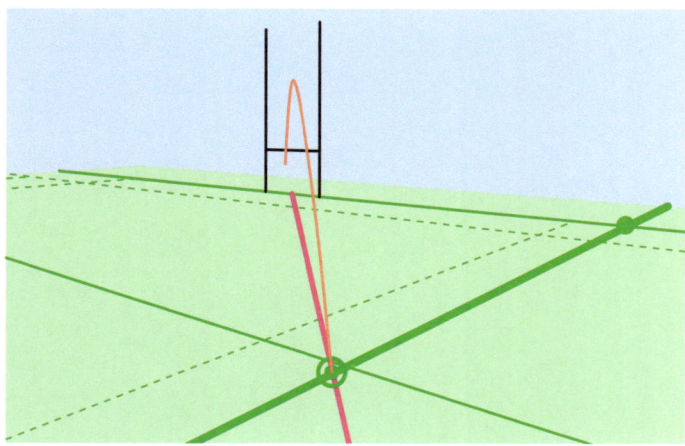

A player's view of the optimal angle

VAULTING VELOCITY

Here's a (slightly dark) question from a pub quiz:

'At what speed would a world-record-breaking pole-vaulter hit the floor, if you removed their crash mat?'

For some important context: in this pub quiz, calculators were explicitly allowed, and the answer needed to be in metres per second, to the nearest whole number. The question prompted significant outrage from many players around the room:

'How can we even answer that!?'
'It depends on how they jump, surely?'
'It depends on so many unknown things – the length of their pole, their run up, etc.'
'What is the world record anyway?'
'Wait, how does gravity work?'

But the answer doesn't depend on any of those unknowns – you just need some basic facts about how gravity works, and an approximate general knowledge of the world-record height for the pole-vault. (In February 2025, it was 6.27m – though current record holder Armand Duplantis shows a remarkable capacity to keep breaking his own record, and doubtless will have done so again more than once by the time you read this.)

Acceleration due to gravity (or how quickly falling things speed up) is often denoted using the letter g, and quoted as a standard value of about 9.8 metres per second per second. Acceleration (a) is related to velocity (final velocity v, and initial velocity u), and displacement (or distance moved, s) by the following equation:

$v^2 = u^2 + 2as$

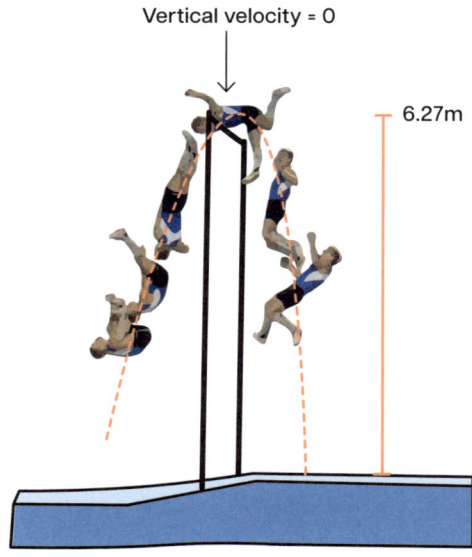

Vertical velocity = 0

6.27m

At the peak of a pole-vaulter's jump – which, in the case of the world record, is roughly when they are at a height of 6.27m – their vertical velocity is 0, as they've stopped going up.

If we use this equation for the second half of the jump, from the peak back to the floor, the displacement s is 6.27m, and the initial velocity is 0, the acceleration is, as always, $9.8m/s^2$. The final velocity, which is what we want to find, is v.

$v^2 = 0 + (2 \times 9.8 \times 6.27)$

$v^2 = 122.982$

$v = 11.09\text{m/s}$

So 11 metres per second seems a reasonable answer. This is close to 25mph, if that helps your intuition. Either way, you definitely don't want to hit something hard (such as the ground) at that speed – which is why crash mats exist.

It's hard to be confident that 11m/s is *exactly* correct, because there are a lot of other possible factors that affect the answer. To name a few: air resistance (friction from the air slows you down a bit as you fall), the fact that the vaulter might not have been exactly at 6.27m (that's the height of the bar, not the vaulter), and the fact that a pole-vaulter also has a bit of horizontal (forward) velocity from their run up (which we've ignored).

But how precise is an answer like this? And where did that magical equation come from?

SUVAT

Precision in calculations is a big difference between textbook answers and the messy real world of predicting things. At school, the mathematical topic of 'upper and lower bounds' for calculations can come across as the most monumentally dull and meaningless topic – 'Why are we bothering with this when I've got the exact answer on my calculator?'

But when modelling the real world, no answer is ever exact, and how far the real answer *might* be from what you calculated becomes desperately important – you might be answering a question like 'how thick should that crash mat be, to make sure the pole-vaulters are safe?'

For example, let's take one of the possible factors that would change our answer – the actual height the pole-vaulter reached. If the bar was at 6.27m, then the pole-vaulter must have gone a bit higher – but this doesn't change the final velocity very much. The graph of the calculation we just did

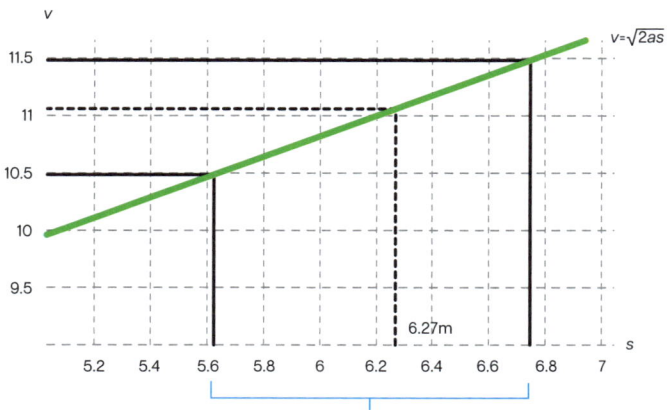

$v=\sqrt{2as}$

6.27m

Any height in this region would give a velocity that rounds to 11m/s

$(v=\sqrt{2as}$, which is a rearrangement of the initial formula, with $u = 0$) shows that any height between about 5.6m and 6.7m would also round to 11m/s. (This graph is not a straight line, even though it looks like one when zoomed in this close.)

The equations we use for situations with constant acceleration (sometimes called the SUVAT equations, because they involve those five letters), are part of a useful mathematical model. These five equations connect initial velocity (u), final velocity (v), displacement (s), time (t) and acceleration (a), under the one big assumption that the acceleration does not change during the situation you're modelling.

THE FIVE SUVAT EQUATIONS

$$v = u + at$$

$$a = \frac{1}{2}(u + v)t$$

$$s = ut + \frac{1}{2}at^2$$

$$s = vt - \frac{1}{2}at^2$$

$$v^2 = u^2 + 2as$$

While that assumption is very rarely exactly true, it's very often close enough to help us model the real world with surprising simplicity. When dealing with anything falling or thrown, that object has a constant acceleration due to gravity, so these equations are useful if we have any questions about sports involving jumping or throwing.

The last equation here is the one we used for the pole-vaulter question.

HOW HIGH CAN YOU THROW?

What about if you want to know how high something gets thrown? Imagine you have video footage of a claimed record-breaking high throw of a tennis ball. It's a fiddly thing to measure directly – having a tape measure in the right place is difficult, let alone getting in the right position to read off the measurement.

On the other hand, if we know how long the ball stays in the air, we can use our SUVAT equations for a calculation. From the video, note the time when it reaches its peak, and then again when it hits the ground – let's say it takes 1.9 seconds between those two points. That time from peak to ground can be used in the third equation to calculate the height.

The velocity at the top of the throw is 0 – that's our initial velocity u (and that's why it's useful to choose to measure from that peak moment); the acceleration a is still $9.8m/s^2$, and t is the time from our video (1.9s). Using these values in the SUVAT equation $s = ut + \frac{1}{2} at^2$, we get:

$s = (0 \times 1.9) + (0.5 \times 9.8 \times 1.9^2)$

$s = 17.68m$ (to 2 decimal places)

If you can get a tennis ball to go that high, you are doing pretty well.

Incidentally, if your object is launched from ground level, you could time the total time from launch to impact with the ground again, and halve it, to get your time from peak to landing. A throw like this will be symmetrical in time, so it'll take the same time to go up as it does to come down.

As always, we've made some assumptions that might affect this answer. Air resistance will change the acceleration very slightly, so the ball is likely to have travelled slightly less far than our calculation (our answer should be an upper bound on the actual height).

An error in timings will affect your answer too: modern videos are often filmed at 30 frames per second, so if you can identify the particular frame when the ball appears to stop moving at the peak, and the frame when it appears to hit the ground, then your timing should be be accurate to within 1/30 of a second. But an error like that means your quoted time might actually be between about 1.87s and 1.93s, which leads to heights of 17.07m and 18.3m respectively – and maybe that's more of a difference than you expected. All told, it would seem sensible to claim a height of at most 17m based on this evidence.

OLYMPIC EQUATIONS

The SUVAT equations end up governing the rough behaviour of anything falling under gravity, so when recent BBC Olympics coverage started including angle and velocity information on some field events, maths teachers everywhere got excited. We can now see just how much more effective an Olympic-level javelin throw is than a ball.

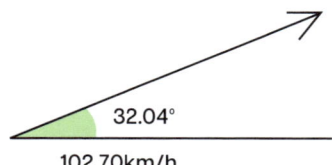

32.04°

102.70km/h

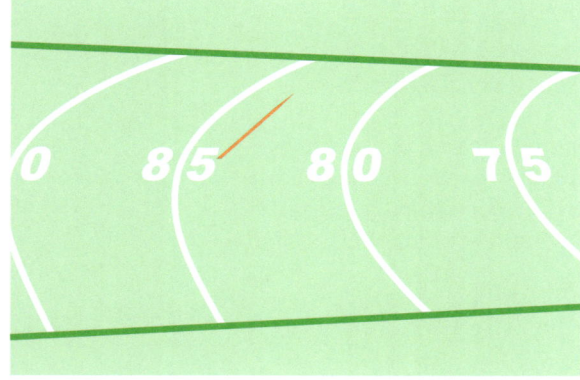

For that real angle and speed given, and taking an average athlete's height into account, the calculations suggest that throw should have gone about 77m.

The actual throw was measured at 84.7m – this is entirely reasonable, since the whole reason javelins are used for throwing is that they fly further than a simple ball, mainly because of their aerodynamics. It's not just that they cut through the air nicely – they also generate lift, like an aircraft wing.

Analysis of many sports is often done using SUVAT equations. If more accurate modelling is needed then these are upgraded to models which take account of the more complicated situations (such as variable acceleration, which means using differential equations – for more on this, see the problem posed on page 124).

Mathematical modelling has helped us simulate many situations, sporting and otherwise, but it's always important to check your assumptions. We don't want parachutists to ignore air resistance, do we?

HOW DOES MATHS HELP FOOTBALL TEAMS PLAY BETTER?

We've already seen many ways in which maths is involved in sport – from the physics of flying objects to the detailed statistics recorded about sporting achievements. But within the sport of football (soccer, to some) we can use maths in even more ingenious ways to get an edge.

For example, in 2010 a team of mathematicians at Queen Mary University in London undertook a detailed mathematical analysis of passing within football teams, for that year's FIFA World Cup. As we saw on page 11, graphs are a great way to model the relationships between things – and we can use them to analyse more complex relationships, like passes between players during a game.

If each player on a football team is a vertex in a graph, we can draw an edge between each pair of players to represent a pass between those two players. A **weighted graph** is one in which each edge of the graph has a number attached to it – and we could use the number to indicate how frequently during a game the ball is passed between those two players.

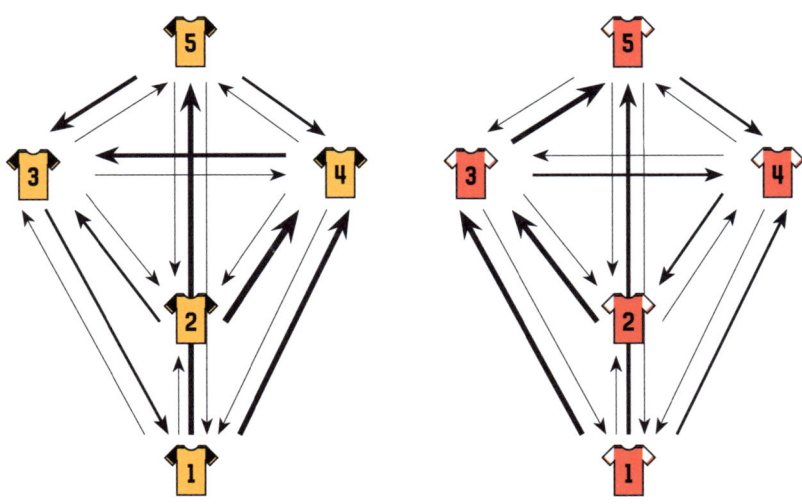

The researchers modelled the teams participating in the World Cup, and identified patterns in the way they passed the ball. For each team, they created a graph, with a line between each pair of players, whose weight – represented by the line thickness – corresponded to how many times those players passed to each other during the game.

These graphs allowed the researchers to analyse which players had the most passes connecting them – if there's a player involved in a large proportion of passes (like our well-connected nodes in the examples on page 12), their removal would have a much bigger impact on the team's passing overall than someone else's removal.

This means that teams with a more even distribution of passes – rather than one or two players doing most of the passing – are more robust to the disruption that might occur if a player was injured or given a red card. The same kind of mathematics is used to analyse computer networks and power grids, to make them more robust to disruption – but, in this case, it tells us a lot about how the team interacts.

WHAT IS SABERMETRICS?

The Society for American Baseball Research (SABR) was founded in the 1970s, and it became a pioneer in the use of statistics to measure sporting performance – first in baseball, but then more widely across other sports. As a result, the term sabermetrics (originally SABRmetrics) has become a blanket term for general sports analytics, and is now very big business indeed.

Huge amounts of money change hands every year when top teams, in football and in many other professional sports, buy and sell players. Sabermetrics is often credited as a way for teams to try to find players who are undervalued: where their transfer cost is low, but the value they'd bring to the team is higher than you might expect.

Determining how valuable a player is can be complex. The cost can depend on a variety of factors, including how well a player has performed at their current team, their age, and sometimes even how many other teams are also interested.

But while simple metrics like 'goals scored in a season' might give a rough indication of how good a player is, in team sports like baseball – and football – there's more to it than that.

The way a player interacts with the rest of the team, and how much they contribute to the game in other ways, as well as factors like how the players might improve in future, are all stronger determinants of how well the team will do overall – and players who score highly in these areas, but don't score so well against more traditional metrics, will potentially be available for a team to purchase more cheaply.

Sabermetrics was developed by analysing a range of different statistics – and baseball is one sport where many different aspects of a game are measured. Batting players are assigned a 'batting average' (what proportion of the time they hit a ball when it's pitched at them, often given as a percentage), and this was historically a major measure of ability.

But baseball fans also keep an eye on other stats, including things like the 'slugging percentage', which records how much movement around the bases results from the player's hits, relative to the number of balls pitched (annoyingly, this not a percentage, but an average, ranging from 0 to 4). Similar metrics also exist for pitchers and fielders.

Sabermetrics brings in all of this data, and tries to analyse how it impacts on a team's performance more generally, allowing teams to predict potential future success. The technique often employs computer programs to process the data – pioneers of the technique used FORTRAN and D-BASE systems, which were the latest technology in the 1970s.

One of the famous success stories among the first teams to use the technique is the California team Oakland Athletics, as detailed in the 2011 film *Moneyball*. In 2002 their team won 20 games in a row, despite having only a small team-player budget.

Could these techniques be used by football teams too? They already have been – Liverpool's manager Jürgen Klopp used the technique in the 2018–2019 season. In that season,

Liverpool lost just one of their 38 regular season games (they lost 12 in the previous season), and went on to win the Champions League that year.

This isn't the only example. Billy Beane, former manager of Oakland Athletics – portrayed in the movie by Brad Pitt – now provides football performance and analytics advice to Dutch team AS Alkmaar. Other major clubs have hired high-profile data analysts too.

HASHTAG GOALS

It's not surprising that data analysis is useful in sports, given the sheer quantity of data generated during games. Modern technology allows us to measure and track hundreds of statistics and facts about a game and the players in it, and in football the analysis often includes reference to the concept of 'expected goals', sometimes denoted 'xG'.

This is a relatively new statistical measure for tracking a football team's performance. The idea is that, given a particular passage of play in a football game, there is some chance that the end result will be a goal. If that chance can be quantified, then numerical analysis of the best way to generate goal scoring chances becomes possible.

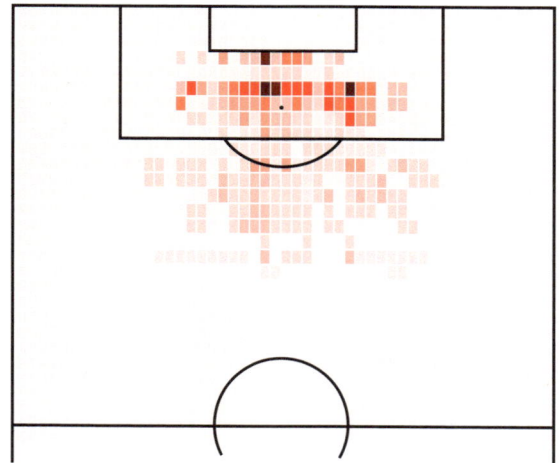

Heat map of areas likely to result in a goal – positions with darker shading are more likely to result in a goal

For example, if you freeze time at the moment a striker is about to take a shot, there will be a number of factors that affect the likelihood of a goal, including the distance from the player to the goal, the shot angle, and how close the nearest defender is.

These factors (among others) are combined to give a probability that a given play will result in a goal – for example, an xG of 0.4 on a particular attempt means that, four times out of ten, taking the same shot, it'll go in the goal. These calculations are based on huge amounts of historical data, and give an idea of not only which parts of the pitch you're most likely to score from, but also where you need to send your defenders in order to prevent your opponents from scoring.

SO...

...how can maths help football teams play better? By allowing them to make good decisions about which players to buy, when to pass, and when to shoot and score.

CAN MATHS MAKE YOU MORE EFFICIENT?

In this modern age, we're all under pressure to get things done quickly and efficiently. And while it's probably not helpful to get too stressed about it, there are ways that maths can help us do things more quickly, simply by applying a bit of logic.

A good example comes in the form of a classic maths puzzle involving a toasting grill, and three slices of bread. You want to toast all three slices on both sides, but the grill lets you put only two slices in at a time, and will toast only the top face of the two pieces you put in. How long (how many grill runs) will it take to toast all three slices?

If you just toast ahead without thinking, you may well end up taking four runs: one run to do the top of A and B, one to do the bottom of A and B, and then you have to do C in two runs, one for the top and one for the bottom.

There is a better way – and the key insight here is that slice C caused the problem, because it was left to the end and you can't do both sides at once.

Labelling the sides of the bread as 1 and 2, if you use the following sequence you'll be done in three runs of the grill:

A1 & B1

A2 & C1

B2 & C2

This puzzle is a simple example of the way mathematical thinking can help us to save time. By knowing which activities take which amounts of time, which can be done at the same time as each other ('parallelized') and which ones need another task to be completed first (called a 'dependency'), we can figure out the most efficient order to do things in.

The mathematical method of **critical path analysis** is widely used in business and industry to optimize projects and workflows. It involves collecting information about tasks and dependencies, and arranging them into an order that works.

The premise will be familiar to anyone who's ever cooked a complicated dinner for a large group of people – you need to boil the potatoes before you can roast them, and the vegetables might not fit in the oven until the joint comes out to rest, and so on.

A diagram called a **Gantt chart** can be used to lay out all the activities. Each row represents a particular task and indicates the resource it will take up (either a member of staff in the case of a business project, or a piece of kitchen equipment if you're cooking dinner). A vertical line from the end of one to the start of another indicates the second task can't start until the other one finishes.

Time:	1–1.30pm	1.30–2pm	2–2.30pm	2.30–3pm
Prep roast	�--			
Cook roast in oven		�--	�--	
Peel potatoes		�--		
Boil potatoes			�--	
Mash potatoes				�--
Prep veg			�--	
Roast veg				�--

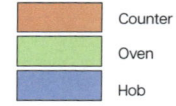

Counter
Oven
Hob

Gantt chart for a roast dinner. The thicker vertical lines indicate that one task can't start until another one has finished. Here, it's clear that the counter isn't being used for more than one task at the same time.

By seeing clearly which things can happen before and after which others, and which can happen concurrently, we can slide all the pieces across to mean there's as little wasted time as possible, and reduce the overall time it takes so it's maximally efficient.

Here's a similar chart for our toast problem:

	Grill run 1	Grill run 2	Grill run 3	Grill run 4
Toast A1	Left			
Toast A2		Left		
Toast B1	Right			
Toast B2		Right		
Toast C1			Left	
Toast C2				Left

■ Left slot on grill

■ Right slot on grill

	Grill run 1	Grill run 2	Grill run 3
Toast A1	Left		
Toast A2		Left	
Toast B1	Right		
Toast B2			Right
Toast C1		Right	
Toast C2			Left

Since there are two slots on the grill, we can have at most two tasks occurring at once; we can also toast only one side of each slice at a time. In the chart, no column has more than two entries, and each pair of rows (representing a single slice of toast) has no more than one entry in it at a time.

On a more day-to-day level, we can use this same kind of thinking to optimize getting around. Imagine you're going to the station to catch a train. You find you'll be there a little early, so will have some time to wait on the platform. How can you make use of this slack time to be more efficient?

Well, if you know the layout of the station at the other end of the journey, you might be able to make a good decision about where to get on the train to minimize the amount of walking you'll need to do at the other end. For example, if you know your destination station has the exit at the far end of the platform in the direction of travel, you can do some of the necessary walking while you've got time to kill at your starting platform. The train won't arrive any earlier, and it's the same amount of walking overall, but you'll already be where you want to go when you get off.

Some transport directions apps, including for complex transport networks like the London Underground, have started including instructions in their directions about where's the best place to wait to board a train so that when you get off you'll be as close as possible to the exit. You might only save a minute or two, but it all adds up!

There are many ways in which mathematical techniques optimize our everyday life, sometimes even without us realizing it. Every time you ask your phone app for directions, there's some serious mathematics happening in the background in order to offer you the most efficient route.

Whether it's car travel, public transport, cycling routes or a walking route, mapping applications rely on the mathematics of graph theory (see page 11) to find efficient paths between two points. They are almost always applying an algorithm called Dijkstra's algorithm, with some customizations, which can find you a 'shortest path' between two nodes on a graph.

SO...

...can maths make you more efficient?
If you use it to optimize your life, then in many ways it certainly can.

WHAT ARE YOU PLAYING AT?

Let's play the 'Fifteen Game'. Write the numbers 1 to 9 individually on small pieces of paper, and we'll take turns to pick one number each to hold on to. Once we have collected at least three numbers each, we'll start checking for a win: the winner is the first person to have exactly three numbers in their collection which add up to 15.

For example, if I picked up the 4, the 3 and the 8, I'd have a set of three numbers that add to 15. (If I have just the 9 and 6, that doesn't count as a win, because even though they add up to 15, there's only two of them.)

PLAYER 1 2 5 7

PLAYER 2 6 9 3

Neither player has won. Player 2 has 9+6=15, but this doesn't count.

PLAYER 1 4 3 8

PLAYER 2 2 5

Player 1 wins (4+3+8=15)

What would your strategy for this game be? Careful analysis at each stage? Wild guessing? It's a difficult one to plan for, since your opponent's moves will make a big difference to the outcome. See if you can find someone who's willing to play against you, and see what happens.

You might be wondering – given the title of this section – how you could possibly give yourself an advantage in a game like this. Well, depending on how you interpreted the first instruction, you might find that the game is significantly easier than you first thought.

If you wrote down the numbers 1 to 9 in a line, that won't help you much. But if instead you wrote them in a square, in a layout that has a top row of 2-7-6, a second row of 9-5-1, and a bottom row of 4-3-8, this suddenly becomes another thing entirely. In fact, it literally is another game entirely.

2	7	6
9	5	1
4	3	8

Those three rows each contain a valid winning set for the game. You'll also find that the columns of your 3-by-3 grid are winning sets: 2-9-4, 7-5-3 and 6-1-8. And, not only that, but the two diagonals of this grid will also get you a win: 2-5-8 and 4-5-6. And these are all the possible sets of three numbers that add to 15.

If you can make sure that on your turns you pick three squares that form a row, column or diagonal in the grid, before your opponent manages to do the same, you'll have won the game. And if that sounds familiar, it should be – this is exactly a description of the game of Tic-Tac-Toe, or Noughts and Crosses. We saw earlier how mathematicians analyse games like this (see page 47), and this is probably a familiar enough game that you already have a good strategy for how to play – which you can then apply to this new game.

This Fifteen Game and Noughts and Crosses are classic examples of **equivalent games**. The mathematical structure is exactly the same, although the experience of playing them, for someone unaware of the link, can feel very different. Incidentally, the grid we suggested is also an example of a **magic square** (all columns, rows and diagonals add to the same total) and is an interesting structure in its own right.

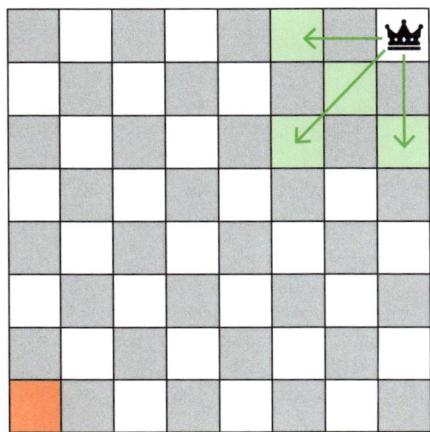

THE QUEEN'S GAME

Here's another fun game you can play: it involves a chessboard, and a single Queen piece. In this game, your moves can be a horizontal move to the left, or a vertical move downwards, or a diagonal move down and left. Unlike in Chess, here the Queen can't move to the right or up, but can only proceed down and/or left.

Players take turns to move the Queen, and the winner is the first person to get to the bottom left square. Like many turn-taking games, there's an advantage for the player who gets to move first – so we'll allow the other player to choose where on the board the Queen starts from.

Your instinct might be to put the Queen as far away from the finish as possible, to prevent your opponent from getting there quickly. But bear in mind that putting the Queen in the diagonally opposite corner would be a terrible choice – your opponent could make a single diagonal move all the way there.

Where would you like to choose as the starting square? If you can find or sketch a chessboard, you might be able to spot some squares you definitely don't want to start from, since they'll be places your opponent can immediately move to the finish from. But these are squares you definitely want to be on when it's your turn – so you might start to build up some idea of what's a good strategy while playing the game too.

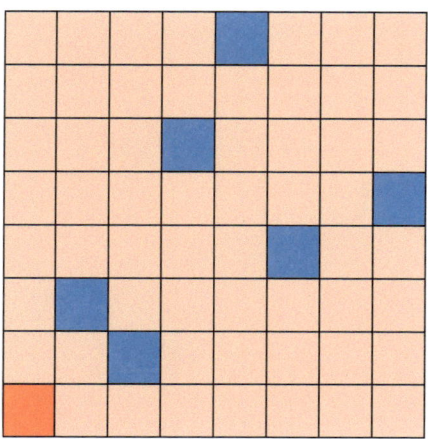

The cold positions are shaded in blue; all other positions are hot

In fact, in this game there are certain squares that are good to be on when it's your turn, often called 'hot' positions, and others where it's bad to start from, or 'cold' positions. For example, anywhere on the diagonal of the board is hot, as are any squares along the bottom or left edges of the board. But a space that's one row up from the bottom and two spaces in from the left is a cold position – since from there, whatever move the player makes, they'll put their opponent into a hot position. So if you can move your opponent into a cold position, you'll be able to win.

The standard 8-by-8 chessboard has six cold positions, including our one-up-two-across position and its mirror image, as well as four others. From these spaces, whatever move your opponent makes, they'll put you in a position to head either straight to the winning space, or to another cold position.

Much like our Fifteen Game, this one (sometimes called **Wythoff's game**) has another, much simpler-seeming game that it's exactly equivalent to. Imagine you have two piles of stones. We can play a game where we take it in turns to take any number of stones from one pile or the other, or the same number of stones from both, and the winner is whoever takes the last stone from the table. If you're going second, you can choose how many stones go in each pile to start with. How many would you like?

Putting the same number in both would be a terrible move, since then your opponent could take both piles and...wait a minute! This is the same situation as before – taking stones from one pile corresponds to moving across the board, taking from the other is like moving down, and taking the same number from each pile is a diagonal move.

This game of stones, known as **two-pile Nim**, is a variant on the Nim game we saw on page 44 – a game which has a long history of mathematical analysis of the best possible move. The two-pile variant requires a bit more careful thinking than the single-pile version, and mathematically it's the same as our earlier Queen's gambit – so the same strategies will work for both.

One way to demonstrate these are equivalent is to describe any position in each game by a pair of numbers. If you label the columns and rows of the chessboard from 0 to 7, then a Queen in the cold position in the 6th column across, 4th row up would be at position (5,3), and that would be exactly the same position in the piles game when there's a pile of 5 and a pile of 3. Position (0,0) corresponds to the winning bottom left square, and to the position when you've just removed the last stones.

This is a good example of the powerful way mathematics can give us tools to take on new challenges. In this case, translating one game into an equivalent game can dramatically change your intuition and insight. This technique can help us analyse all sorts of things, from simple games to cracking unsolved problems. Being able to identify the fundamental structure something is based on, and match it to an existing structure we understand, means we can apply the tools we have in one situation to a totally different one.

SO...

...**what are you really playing at?** With mathematics, the answer might not be what you thought it was – you might be playing a different game altogether.

THE CAR CRASH PROBLEM

Drivers often rely on intuition to make quick decisions in driving situations. How's your intuition? Try it out here.

Imagine you are driving – in the blue car – on a motorway, at the legal limit in the UK of 70mph (just over 110km/h). You notice a speeding car overtaking you (the red car), going at 100mph (a little over 160km/h).

Now imagine that at the moment when the cars are exactly level, you both notice a hazard on the road ahead – a fallen tree, for example, blocking the whole width of the road – and both fully slam on the brakes.

Thankfully, you manage to bring your blue car to a stop with millimetres to spare. What will happen to the red car?

In particular, since it's fairly obvious that it will crash, *at what speed will the red car hit the tree?*

70mph   100mph

Check your gut-reaction guess at an answer before attempting any calculations (or checking the solution). Then you can try to find a more precise solution, or read on.

Like all attempts to model the real world, you're allowed to assume whatever you like to keep it simple enough to get started – but remember to check what effect changing those assumptions will have later.

In this case you could assume:

- that both drivers react instantly (so we can ignore reaction times)
- that both cars are identical (in particular, assume both cars brake at the same rate, which remains constant)
- that they don't skid or do anything complicated while slowing down.

With those assumptions at least, there is a fairly definitive answer. We'll discuss what happens if you want to start including these extra complications later.

The answer to this puzzle, with the assumptions mentioned, is that:

Red will hit the tree travelling at about 71mph.

IS THAT REALLY RIGHT?

Yes, you read that right – *the red car does not even manage to slow down to the speed at which blue started braking*. This sobering outcome does rely on the assumptions mentioned, but it is justifiable in several ways.

Many people find the outcome counterintuitive, perhaps because our intuition often relies on 'linear thinking' – for example, doubling one thing might involve doubling the other.

Here, your speed is not in a linear relationship with your energy (and hence your stopping distance). The relationship between speed and energy has a squared term in it, and this means our intuition can be less reliable.

Assume red and blue have the same (constant) deceleration. Then we can use the standard 'constant acceleration formulae' from page 168. Here's a reminder of the five variables they involve, and the five equations:

- u = initial velocity

- v = final velocity

- a = acceleration (which will be negative for deceleration)

- t = time

- s = displacement (the distance away from the starting point)

$$v = u + at$$

$$a = \frac{1}{2}(u+v)t$$

$$s = ut + \frac{1}{2}at^2$$

$$s = vt + \frac{1}{2}at^2$$

$$v^2 = u^2 + 2as$$

In this case, we're assuming that a is the same for each car (but we don't know what it is) and also the value for s is identical because the cars are the same distance away from the tree (even though we don't know how far that is). We don't know about t (the time taken to reach the tree) and in any case it will be different for each car (red, going quicker, is going to hit much sooner than blue manages to stop).

The final equation, $v^2 = u^2 + 2as$, then becomes useful – because it doesn't involve the unknown t variable.

For blue, the final velocity is 0 (you have stopped), and the initial velocity is 70. We don't know a or s, but we can write:

$$0^2 = 70^2 + 2as$$

Rearranging gives $2as = -4,900$.

Crucially, even though we don't know a or s, we now know the value of $2as$ is $-4,900$. It is negative because the acceleration is in fact a deceleration.

For red, we want to calculate the final velocity (let's leave it as v), and we know the initial velocity is 100.

$$v^2 = 100^2 + 2as$$

We've already worked out that $2as = -4,900$, so:

$$v^2 = 10,000 - 4,900$$

$$v^2 = 5,100$$

Taking the square root gives:
$v = 71.4\text{mph}$

A simpler calculation can justify the same answer if we assume that braking is the process of removing kinetic energy, and that the brakes do this at the same rate over the distance the brakes are applied. Kinetic energy (KE) is calculated by the formula:

$$KE = \frac{1}{2} mv^2$$

Here m is the mass of the object, and v is the velocity. Assuming the masses of the cars are the same, and that the cars get rid of the same KE when they brake, we can look at the difference in their original kinetic energy to see what red will have left when hitting the tree.

The difference is $m \times (100^2 - 70^2)$, so when blue stops, red will still have a KE of $5,100 \times m$. The 5,100 corresponds to the velocity squared, so square rooting that value, as before, gives 71.4mph.

WHAT ABOUT THE ASSUMPTIONS?

Upgrading the assumptions to be more realistic does not help red. Firstly, humans don't react instantly. If both cars react to the situation after the same reaction time, then red has already moved much closer to the tree than blue. So improving the model to include reaction times will make the final crash speed for red even worse.

The other assumption worth challenging is the one about the braking rate being constant, and identical for both cars. It is almost certain that in reality this deceleration will change over the course of the braking period. These things require the mathematical model to get more technical – the acceleration is no longer assumed to be constant, so the 'constant acceleration equations' no longer apply. Instead we'd use differential equations (as on page 124) – which are the general mathematical tools we use to describe the rates at which things are changing.

Without going into technicalities, you can argue that things tend to slow down more quickly if they are moving more quickly, because frictional forces are much greater when moving at higher speeds. This will tend to reduce the predicted crash speed for red a little bit – but it does not go nearly far enough to stop the outcome being tragic.

The moral of the story? Your intuition about speed is probably not as good as you'd like. Pay attention to speed limits!

INDEX

ABOUT THE AUTHORS

BEN SPARKS is a freelance mathematician, teacher, musician and public speaker who delivers maths talks and workshops worldwide. He also works with the University of Bath and the maths education charity MEI. He has co-authored A-level Further Mathematics textbooks and popular maths books, including *100 Ideas in 100 Words* for the Science Museum. As a presenter and content creator, Ben has contributed to popular educational YouTube channels including Numberphile (4.55 million subscribers), where his videos have been watched over 20 million times.

DR KATIE STECKLES is a mathematician and prominent science communicator. She delivers maths talks, workshops and events, and regularly appears on platforms like YouTube, TV (including the BBC, Channel 4 and the Discovery Channel in the US), as well as BBC radio. Katie has written for *New Scientist* magazine and has co-authored books such as *The Math of a Milkshake* and *Short Cuts: Maths*. In 2016, she won the Joshua Phillips Award for Innovation in Science Engagement and was the London Mathematical Society's popular maths lecturer in 2018. In 2025, Katie won the Beetlestone Award for Leadership and Legacy in Science Communication.